SpringerBriefs in Applied Sciences and Technology

Thermal Engineering and Applied Science

Series Editor

Francis A. Kulacki, Department of Mechanical Engineering, University of Minnesota, Minneapolis, MN, USA

More information about this series at http://www.springer.com/series/10305

Sujoy Kumar Saha • Hrishiraj Ranjan
Madhu Sruthi Emani • Anand Kumar Bharti

Insert Devices and Integral Roughness in Heat Transfer Enhancement

 Springer

Sujoy Kumar Saha
Mechanical Engineering Department
Indian Institute of Engineering Science
and Technology, Shibpur
Howrah, West Bengal, India

Hrishiraj Ranjan
Mechanical Engineering Department
Indian Institute of Engineering Science
and Technology, Shibpur
Howrah, West Bengal, India

Madhu Sruthi Emani
Mechanical Engineering Department
Indian Institute of Engineering Science
and Technology, Shibpur
Howrah, West Bengal, India

Anand Kumar Bharti
Mechanical Engineering Department
Indian Institute of Engineering Science
and Technology, Shibpur
Howrah, West Bengal, India

ISSN 2191-530X   ISSN 2191-5318   (electronic)
SpringerBriefs in Applied Sciences and Technology
ISSN 2193-2530   ISSN 2193-2549   (electronic)
SpringerBriefs in Thermal Engineering and Applied Science
ISBN 978-3-030-20775-5   ISBN 978-3-030-20776-2   (eBook)
https://doi.org/10.1007/978-3-030-20776-2

This Springer imprint is published by the registered company Springer Nature Switzerland AG
The registered company address is: Gewerbestrasse 11, 6330 Cham, Switzerland

Contents

Nomenclature

Δp	Fluid static pressure drop on one side of a heat exchanger core, Pa or lbf/ft^2
ΔT_{fs}	Temperature difference between fluid and surface, K or °F
ΔT_i	Temperature difference between hot and cold inlet fluids, K or °F
ΔT_{lm}	Log-mean temperature difference between fluid and surface, K or °F
A	Heat transfer surface area, m^2 or ft^2
A_c	Flow cross-sectional area in minimum flow area, m^2 or ft^2
A_{cj}	Cross-sectional area for tangential injection, m^2 or ft^2
C	Capacity rate $(=WC_p)$, kJ/kg or Btu/h °F
C_p	Specific heat of fluid at constant pressure, J/kg K or Btu/lbm sf
D_h	Hydraulic diameter of flow passages, $4LA_c/A$, m or ft
D_V	Volumetric hydraulic diameter, 4 × void volume/total surface area, m or ft
d_c	Outside diameter of wire coil used to make insert, m or ft
d_i	Tube inside diameter, or diameter to the base of internal fins or roughness, m or ft
d_o	Tube outside diameter, fin root diameter for a finned tube, m or ft
e	Wire diameter for wire coil insert, m or ft
f	Fanning friction factor, $\Delta P_f d_i/2LG^2$, dimensionless
f_{Dh}	Fanning friction factor based on $D_h \Delta P_f \cdot D_h/2LG^2$, dimensionless
f_{sw}	Fanning friction factor for swirl flow, $\Delta P_f d_i/2L(G_{SW})^2$, dimensionless
G	Mass velocity based on the minimum flow area, kg/m^2 s or lbm/ft^2 s
Gr_d	Grashof number $= g_r \beta \Delta T d^3/v^2$, dimensionless
Gr_{Dh}	Grashof number $= g_r \beta \Delta T D_h^3/v^2$, dimensionless
G_{SW}	Mass velocity based on u; and the minimum flow area, kg/m^2 s or lbm/ft^2 s
Gz	Graetz number $= \pi d_i RePr/4L = \pi/4L_d$, dimensionless
g	Gravitational acceleration 9.806 m/s^2 or 32.17 ft/s^2
g_r	Radial acceleration, m/s^2 or ft/s^2
H	Length for 180° revolution of twisted tape, m or ft
h	Heat transfer coefficient based on A, h (local value), W/m^2 K or Btu/h ft^2 °F
j	Colburn factor $= StPr^{2/3}$, dimensionless

k	Thermal conductivity of fluid, W/m K or Btu/h ft °F
L	Fluid flow length, m or ft
L_S	Swirl length of twisted tape ($=L/\cos a$), m or ft
L°	$Lld/Re_d Pr$, dimensionless
LMTD	Log-mean temperature difference, K or °F
m	Number of twisted tape modules, dimensionless
M_T	Axial momentum, kg m/s^2 or lbm ft/s^2
M_t	Tangential injected momentum ($= W?lpA_c$), kg m/s^2 or lbm ft/s^2
Nu_{Dh}	Nusselt number based on D_h ($= hD_h/k$), dimensionless
Nu_{Dv}	Nusselt number based on $D1$ ($= hD_v/k$), dimensionless
Nu_d	Nusselt number based on d, ($= hd_i/k$), dimensionless
NTU	Number of heat transfer units, dimensionless
P	Fluid pumping power, W or hp
Pr	Prandtl number $= cP\mu Jk$, dimensionless
P	Axial pitch of wire or roughness elements, m or ft
P_r	$2H$, m or ft
Q	Heat transfer rate in the exchanger, W or Btu/h
r	Tube or annulus radius, m or ft
Ra	Rayleigh number $= Gr\, Pr$, dimensionless
R_f	Tube-side fouling resistance, m^2 K/W or ft^2 h °F/Btu
Re_d	Reynolds number based on the tube diameter $= Gd/\mu$, dimensionless
Re_{Dh}	Reynolds number based on the hydraulic diameter $= GDj\mu$, dimensionless
Re_{Dv}	Reynolds number based on D_v, dimensionless
Re_s	Equivalent smooth tube Reynolds number $= (f!J;,)\, Re$, dimensionless
Re_{sw}	Swirl Reynolds number $= d;u,wpl\mu$, dimensionless
Re_N	Reynolds number based on axial distance ($= u = x/v$), dimensionless
S_t	Longitudinal tube or element pitch, m or ft
St	Stanton number $= hiGc$, dimensionless
S_w	Swirl number, $Re_{sw}y^{-1/2}$, dimensionless
t	Thickness of fin or twisted tape, m or ft
U	Overall heat transfer coefficient, W/m^2 K or Btu/h ft^2 °F
u_c	Fluid mean axial velocity at the minimum free flow area, m/s or ft/s
u_∞	Free stream velocity over flat plate, m/s or ft/s
u_θ	Tangential velocity, m/s or ft/s
$u*$	Friction velocity $= (\tau_w/\rho)^{1/2}$ m/s or ft/s, dimensionless
W	Fluid mass flow rate, kg/s or lbm/s, or width of tape, m or ft
W_t	Tangentially injected fluid mass flow rate, kg/s or lbm/s
x	Cartesian coordinate along the flow direction, m or ft
x_d	$x/d_i Re_d Pr$
y	Twist ratio $= H/d;\, = \pi/(2\tan\alpha)$, dimensionless
y_b	Thickness of viscous influenced fluid layer, m or ft
$y*$	$yu*/v$ dimensionless
z	Dimensionless pitch of segmented inserts (Z/d_i)
Z	Spacing between tape segments

Greek Symbols

α	Heat transfer coefficient or apex angle of the fin or thermal diffusivity
β	Helix angle or volume coefficient of thermal expansion, 1/K or 1/R
δ	Liquid film thickness
δ_L	Thickness of laminar sublayer on flat plate, m or ft
δ_t	Thermal boundary layer thickness, m
Δp	Pressure drop
ΔT	Temperature difference
ε	Permittivity
ε	Porosity, dimensionless
ε_h	Eddy diffusivity for heat, m^2/s or ft^2/s
ε_m	Eddy diffusivity for momentum, m^2/s or ft^2/s
η_f	Fin efficiency or temperature effectiveness of the fin, dimensionless
λ	Parameter in empirical correlations
μ	Dynamic viscosity
ρ	Density
ρ_m	Average density
σ	Area ratio A_{fr}/A_c, surface tension
τ_o	Apparent wall shear stress, Pa or lbf/ft^2
τ_w	Wall shear stress, Pa or lbf/ft^2
ν	Kinematic viscosity, m^2/s or ft^2/s
ν_ε	Effective viscosity in turbulent region

Subscripts

ct	Continuous tape
fd	Fully developed flow
FR	Fully rough condition
H	Constant heat flux thermal boundary condition
m	Average value over tube cross section or flow length
p	Plain tube or surface
sg	Sand-grain roughness
st	Segmented tape
T	Constant wall temperature thermal boundary condition
w	Evaluated at wall temperature
x	Local value
δ	Viscous boundary layer thickness

Chapter 1
Introduction

Insert devices are primarily used for single-phase flow, and enhancement techniques increase the tube-side convective heat transfer.

Insert devices involve various geometric forms, and these are inserted in smooth tubes. Performance and initial cost dictate which method will be preferred. Integral internal fins and roughness require deformation of the material on the inside surface of a long tube. The deformation of the inner surface of a tube must be manufactured in a cost-effective manner. For turbulent flow, insert devices are not good. Insert devices are best suited for laminar flow. Nevertheless, the insert devices are good for making the performance of an existing heat exchanger better.

Figure 1.1 shows tape inserts. Twisted tape inserts (Fig. 1.1a) make the flow helically imparting swirl into the flow with the addition of radial velocity and tangential velocity to the main axial velocity along the axis of the channel. The more the swirl intensity is due to tighter twist in the tape, the more the enhancement is. However, the tape inserts seldom have good thermal contact with the tube wall, and the twisted tapes do not act as a fin.

Figure 1.2 shows extended surface inserts. Extended surface inserts are the extruded shapes inserted into the tube; the tube is drawn to provide good thermal contact between the wall and the insert. The insert reduces the hydraulic diameter and acts as an extended surface.

Figure 1.3 shows wire coil inserts. This type of inserts consists of a helical coiled spring that functions as a non-integral roughness.

Figure 1.4 shows mesh or brush inserts. Figure 1.5 shows displaced insert devices. These devices are displaced from the tube wall, and they cause periodic mixing. The concepts shown in Fig. 1.5 have been utilized in the aforementioned insert devices. Work with the above-mentioned devices has by and large been empirical. There are many literatures on these types of inserts.

For turbulent flow, the dominant thermal resistance is very close to the wall, and it is, therefore, more effective to mix the flow in the viscous boundary layer at the wall

© The Author(s), under exclusive license to Springer Nature Switzerland AG 2020
S. K. Saha et al., *Insert Devices and Integral Roughness in Heat Transfer Enhancement*, SpringerBriefs in Applied Sciences and Technology,
https://doi.org/10.1007/978-3-030-20776-2_1

Fig. 1.1 Tape inserts: (**a**) continuous twisted tape, (**b**) segmented twisted tape, (**c**) Kinex mixer, (**d**) helical tape, (**e**) bent strips (Webb and Kim 2005)

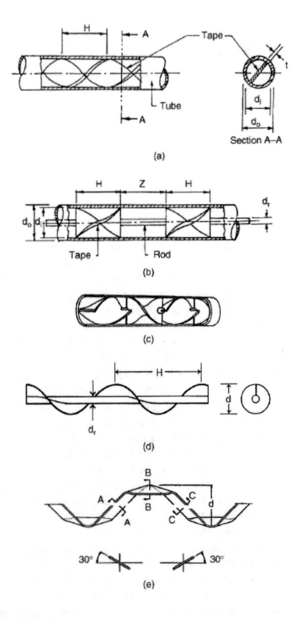

than to mix the gross flow. Integral roughness provides enhancement with higher efficiency index than that provided by wire coil insert.

In the laminar flow regime, the dominant thermal resistance is not limited to thin boundary layer adjacent to the flow. Enhancement devices mixing the gross flow are more effective in laminar flow than in turbulent flow. The extended surface insert is suitable for both laminar flow and turbulent flow. But, it is costly. The twisted tape

Fig. 1.2 Extended surface
inserts: (**a**) extruded insert
(courtesy Wieland-Werke
AG), (**b**) interrupted sheet
(Webb and Kim, 2005)
metal (from Jayaraj et al.
1989)

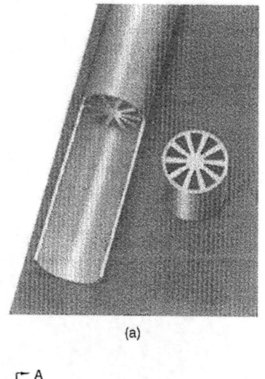

(a)

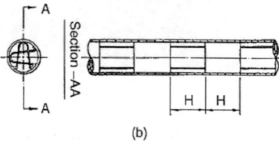

(b)

and mesh inserts are more suited to laminar flow. But, since the tape is in good
thermal contact with the wall, their swirl-generating ability cannot be fully exploited.

Forced convection flow over rough surfaces may be categorized for six different
flow geometries, for example flat plates, circular tubes, non-circular channels in gas
turbine blades, flow normal to circular tubes, annuli having roughness on the outer
surface of the inner tube and longitudinal flow in rod bundles.

Internally roughened tubes and roughness on the water side of evaporators and
condensers are commonly used in refrigeration industry (Webb and Robertson 1988;
Webb 1991; Webb et al. 1984; Jaber et al. 1991). Modern gas turbine blades contain
roughened channels (Fig. 1.6) (Han et al. 1988, 1991; Han (1984); Metzger et al.

Fig. 1.3 Wire coil inserts: (a) wires touching tube wall, (b) wires displaced from tube wall (Webb and Kim 2005)

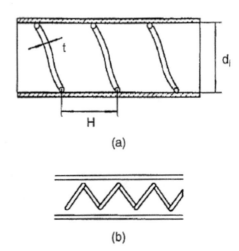

(a)

(b)

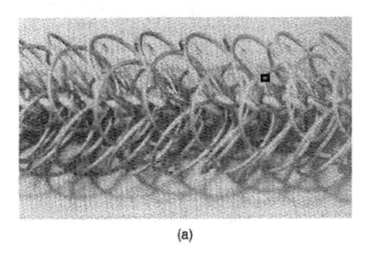

(a)

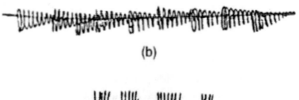

(b)

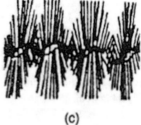

(c)

Fig. 1.4 Mesh or brush inserts: (a) mesh insert, (b) helical coil insert, (c) brush insert (Webb and Kim 2005)

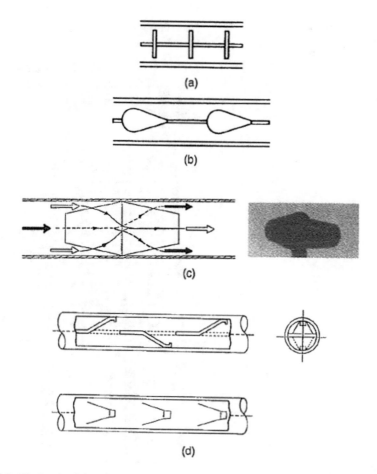

Fig. 1.5 Displaced mixing devices: (**a**) spaced disks, (**b**) spaced streamline shapes, (**c**) flow eversion device (from Maezawa and Lock 1978), (**d**) louvred strip (Webb and Kim 2005)

1983, 1987). Roughened fuel rods are used in gas-cooled nuclear reactors (Dalle-Donne and Meyer 1977).

Roughness geometries are shown in Fig. 1.7. There are three basic roughness families. Key dimensionless variables to define roughness are dimensionless roughness height (e/d), dimensionless roughness spacing (p/e), dimensionless rib width (w/e) and the shape of the roughness element. The ridge-and-groove-type roughness may also be applied at a helix angle (α). A family of geometrically similar roughness, for any specific type of roughness, is possible by changing e/d with constant p/e and w/e; thus virtually unlimited number of specific roughness geometries and sizes is possible. Figure 1.8 shows commercially used roughness geometries.

Heat transfer design methods for integral roughness depend primarily on the design objectives and design constraints. Some information on these aspects may be obtained from Webb and Kim (2005). Some idea on preferred roughness type and

Fig. 1.6 Cooling concepts
of a modern multipass
turbine blade (from Han
et al. 1988)

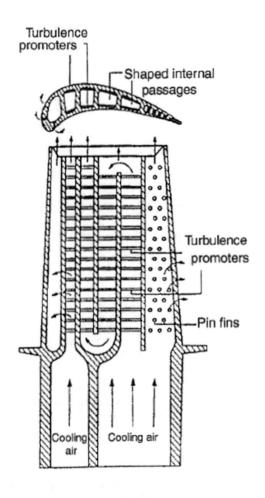

size may also be obtained from Webb and Kim (2005). Webb and Kim (2005) also
discussed general performance characteristics like St and f versus Reynolds number.
Webb (1979), Burck (1970), Webb and Eckert (1972), Edwards (1966), Ye et al.
(1987), Sheriff and Gumley (1966), Ravigururajan and Bergles (1985, 1996), Esen
et al. (1994) and Webb et al. (1971) proposed that all geometrically similar roughness of interest exhibit some qualitative characteristics and give other correlating
methods. Some of the correlations are as follows:

$$\frac{f}{f_s} = \left[1 + \left[\frac{29.1Re_d^{(0.67-0.06p/d_i-0.49\alpha/90)}(e/d_i)^{(1.37-0.157p/d_i)}(p/d_i)^{(-1.66E-6Re_d-0.33\alpha/90)}}{\left(\frac{\alpha}{90}\right)^{(4.59+4.11E-6Re_d-0.15p/d_i)}(1+2.94\sin(90-\theta/n))}\right]^{15/16}\right]^{16/15}$$

$$(1.1)$$

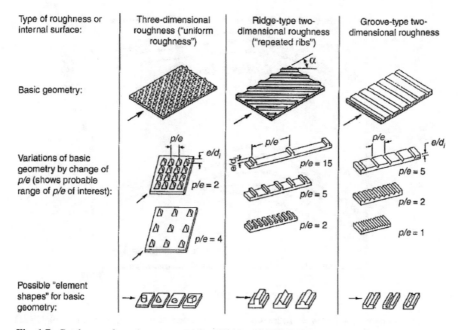

Fig. 1.7 Catalogue of roughness geometries (Webb and Kim 2005)

Fig. 1.8 Illustration of commercially used enhanced tubes: (**a**) helical rib Turbo-Chilf tube, (**b**) corrugated Korodense'Y tube (courtesy of Wolverine Tube, Decatur, AL) (Webb and Kim 2005)

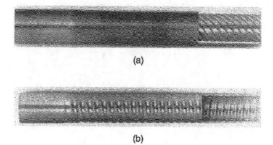

$$\frac{Nu}{Nu_{\mathrm{s}}} = \left[1 + \left[2.64Re_{\mathrm{d}}^{(0.036)}(e/d_{\mathrm{i}})^{(0.212)}(p/d_{\mathrm{i}})^{(-0.21)}\left(\frac{\alpha}{90}\right)^{0.29}Pr^{0.024}\right]^{7}\right]^{1/7} \quad (1.2)$$

Rough surfaces show a significantly different Prandtl number dependency than smooth surfaces (Fig. 1.9). Modified Petukhov (1970) equation for smooth tubes is

$$St = \frac{f/2}{1 + \sqrt{f/2}[\bar{g}(e^{+})Pr^{n} - B(e^{+})]} \quad (1.3)$$

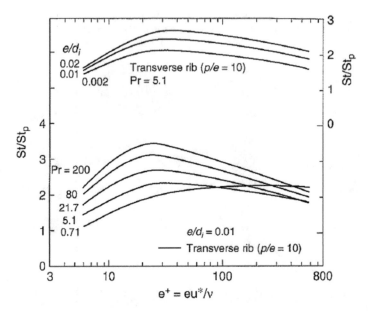

Fig. 1.9 St/St_P vs. $e+$ for $p/e = 10$ transverse rib roughness calculated from Webb et al. (1971) correlation for different Prandtl numbers (from Webb 1979)

At high and/or Pr, $\bar{g}Pr^n \gg B$, the asymptotic value of the above equation may be written as

$$\frac{f/2St - 1}{\sqrt{f/2}} \approx \bar{g}(e^+)Pr^n \tag{1.4}$$

For $Pr \gg 1$,

$$\frac{f/2St - 1}{\sqrt{f/2}} \cong 1.27Pr^{2/3} \tag{1.5}$$

Table 1.1 shows that rough tubes do not have the same $Pr^{2/3}$ exponent as smooth tubes; this observation was made by Withers (1980a, 1980b), Sethumadhavan and Raja Rao (1983, 1986), Dawson and Trass (1972), Brognaux et al. (1997), Dong et al. (2001), Vincente et al. (2002a), Webb et al. (1971) and Dipprey and Sabersky (1963) over a wide Prandtl number range.

Sarma (2003), Hsieh and Huang (2000), Hsieh (2003), Manglik and Bergles (1992), Chang et al. (2007a, b) and Saha and Mallick (2005) reported on heat transfer enhancement by using twisted tape inserts in a tube. Tube with wire coil insert was used to enhance the heat transfer by Naphon (2006). Many investigators Williams et al. (1970), Webb et al. (1971), Han et al. (1978), Zhang et al. (1994), Prasad et al. (2009), Prasad and Saini (1988), Jaurker et al. (2006), Layek et al.

Table 1.1 Prandtl number exponents for rough surfaces (Webb and Kim 2005)

Reference	n	Pr (or Sc)	Roughness type
Dipprey and Sabersky (1963)	0.44	1.2–5.94	Sand grain
Webb et al. (1971)	0.57	0.7–37.6	Ribs (10 < ple < 40)
Sethumadhavan and Raja Rao (1986)	0.55	5.2–32	Corrugated
Sethumadhavan and Raja Rao (1983)	0.55	5.2–32	Wire coil inserts
Dawson and Trass (1972)	0.58	320–4500	Ribs (ple ≃ 3.7)
Withers (1980a)	0.50	2.5–10	Corrugated
Withers (1980b)	0.50	4.7–9.3	Helical rib
Vicente et al. (2002)	0.60	2.5–100	Three-dimensional helical dimple
Brognaux el al. (1997)	0.56–0.57	0.7–7.85	Microfin
Liao et al. (2000)	0.51	2.9–13.8	Three-dimensional protrusion
Dong et al. (2001)	0.55	2–140	Corrugated

(2007), Bhagoria et al. (2002) and Hassan Ridouane and Campo (2008) also investigated the effects of the rib angles, pitch–height ratio and rib groove roughened surfaces on friction factor and heat transfer enhancement. Deshmukh and Vedula (2014), Oni and Paul (2016), Eiamsa-ard and Promvonge (2007), Deshmukh et al. (2016), Chang et al. (2007a, b), Murugesan et al. (2011), Eiamsa-ard et al. (2008), Saravanan et al. (2016) and Mwesigye et al. (2016) worked on heat transfer enhancement by using different types of inserts.

References

Bhagoria JL, Saini JS, Solanki SC (2002) Heat transfer coefficient and friction factor correlation for rectangular solar air heater duct having transverse wedge shaped rib roughness on the absorber plate. J Renew Energy 25:341–369

Brognaux L, Webb RL, Charura LM, Chung BY (1997) Single-phase heat transfer in microfin tubes. J Heat Mass Transf 40:4345–4358

Burck E (1970) The influence of Prandtl number on heat transfer and pressure drop of artificially roughened channels. In: Bergles AE, Webb R (eds) Augmentation of convective heat and mass transfer. ASME, New York

Chang SW, Jan YJ, Liou JS (2007a) Turbulent heat transfer and pressure drop in tube fitted with serrated twisted tape. Int J Therm Sci 46:506–518

Chang SW, Yang TL, Liou JS (2007b) Heat transfer and pressure drop in tube with broken twisted tape insert. Exp Therm Fluid Sci 32(2):489–501

Dalle-Donne M, Meyer L (1977) Turbulent convection heat transfer from rough surfaces with two dimensional rectangular ribs. Int J Heat Mass Transf 20:583–620

Dawson DA, Trass O (1972) Mass transfer at rough surfaces. Int J Heat Mass Transf 15:1317–1336

Deshmukh PW, Vedula RP (2014) Heat transfer and friction factor characteristics of turbulent flow through a circular tube fitted with vortex generator inserts. Int J Heat Mass Transf 79:551–560

Deshmukh PW, Prabhu SV, Vedula RP (2016) Heat transfer enhancement for laminar flow in tubes using curved delta wing vortex generator inserts. Appl Therm Eng 106:1415–1426

Dipprey DF, Sabersky RH (1963) Heat and momentum transfer in smooth and rough tubes at various Prandtl numbers. Int J Heat Mass Transf 6:329–353

Dong Y, Huixiong L, Tingkuan C (2001) Pressure drop, heat transfer and performance of single-phase turbulent flow in spirally corrugated tubes. Exp Therm Fluid Sci 24:131–138

Edwards FJ (1966) The correlation of forced convection heat transfer data from rough surfaces in ducts having different shapes of flow cross section. In: Proceedings of the third international heat transfer conference, vol 1, pp 32–44

Esen EB, Obot NT, Rabas TJ (1994) Enhancement: Part II. The role of transition to turbulent flow. J Enhanc Heat Transf 1:157–167

Eiamsa-ard S, Promvonge P (2007) Heat transfer characteristics in a tube fitted with helical screw-tape with/without core-rod inserts. Int Commun Heat Mass Transf 34:176–185

Eiamsa-ard S, Pethkool S, Thianpong C, Promvonge P (2008) Turbulent flow heat transfer and pressure loss in a double pipe heat exchanger with louvered strip inserts. Int Commun Heat Mass Transf 35:120–129

Han JC (1984) Heat transfer and friction in channels with two opposite rib-roughened walls. J Heat Transf 106:774–781

Han JC, Glicksman LR, Rohsenow WM (1978) An investigation of heat transfer and friction for rib-roughened surfaces. Int J Heat Mass Transf 21:1143–1156

Han JC, Chandra PR, Lau SC (1988) Local heat/mass transfer distributions around sharp 180 deg turns in two-pass smooth and rib-roughened channels. J Heat Transf 110:91–98

Han JC, Zhang YM, Lee CP (1991) Augmented heat transfer in square channels with parallel, crossed, and V-shaped angled ribs. J Heat Transf 113:590–596

Hassan Ridouane EI, Campo A (2008) Heat transfer enhancement of air flowing across grooved channels: joint effects of channel height and groove depth. ASME J Heat Transf 130:1–7

Hsieh (2003) Experimental studies on heat transfer and flow characteristics for turbulent flow of air in a horizontal circular tube with strip type insert

Hsieh and Huang (2000) Experimental studies for heat transfer and pressure drop of laminar flow in horizontal tubes with/without longitudinal inserts

Jaber MH, Webb RL, Stryker P (1991) An experimental investigation of enhanced tubes for steam condensers, ASME Paper 91-HT-5. ASME, New York

Jaurker AR, Saini JS, Gandhi BK (2006) Heat transfer coefficient and friction characteristics of rectangular solar air heater duct using rib-grooved artificial roughness. Int J Solar Energy 80:895–907

Jayaraj D, Masilamani JG, Seetharamu KN (1989) Heat transfer augmentation by tube inserts in heat exchangers, SAE Technical paper 891983, Warrendale, PA

Layek A, Saini JS, Solanki SC (2007) Heat transfer coefficient and friction characteristics of rectangular solar air heater duct using rib-grooved artificial roughness. Int J Heat Mass Transf 50:4845–4854

Liao Q, Zhu X, Xin MD (2000) Augmentation of turbulent convective heat transfer in tubes with three-dimensional internal extended surfaces. J Enhanc Heat Transf 7(3):139–151

Manglik RM, Bergles AE (1992) Heat transfer enhancement and pressure drop in viscous liquid flows in isothermal tubes with twisted-tape inserts. Heat Mass Transf 27(4):249–257

Maezawa S, Lock GSH (1978) Heat transfer inside a tube with a novel promoter. In: Heat transfer 1978. Proceedings of the 6th international heat transfer C011f, vol 2. Hemisphere, Washington, DC, pp 596–600

Metzger DE, Fan CZ, Pennington JW (1983) Heat transfer and flow friction characteristics of very rough transverse ribbed surfaces with and without pin fins. In: Proc. 1983 ASME-JSME thermal engineering conference, vol 1, pp 429–436

Metzger DE, Vedula RP, Breen DD (1987) The effect of rib angle and length on convection heat transfer in rib-roughened triangular ducts. In: Proc. 1987 ASME-JSME thermal engineering conference, vol 3, pp 327–333

Murugesan P, Mayilsamy K, Suresh S, Srinivasan PSS (2011) Heat transfer and pressure drop characteristics in a circular tube fitted with and without V-cut twisted tape insert. Int Commun Heat Mass Transf 38:329–334

Mwesigye A, Bello-Ochende T, Meyer JP (2016) Heat transfer and entropy generation in a parabolic trough receiver with wall-detached twisted tape inserts. Int J Therm Sci 99:238–257

Naphon P (2006) Effect of coil-wire insert on heat transfer enhancement and pressure drop of the horizontal concentric tube. Int Commun Heat Mass Transf 33:753–763

Oni TO, Paul MC (2016) Numerical investigation of heat transfer and fluid flow of water through a circular tube induced with divers' tape inserts. Appl Therm Eng 98:157–168

Petukhov BS (1970) Heat transfer in turbulent pipe flow with variable physical properties. In: Irvine TF, Hartnett JP (eds) Advances in heat transfer, vol 6. Academic Press, New York, pp 504–564

Prasad BN, Saini JS (1988) Effect of artificial roughness on heat transfer and friction factor in a solar air heater. Sol Energy 41(6):555–560

Prasad SB, Saini JS, Singh Krishna M (2009) Investigation of heat transfer and friction characteristics of packed bed solar air heater using wire mesh as packing material. Int J Solar Energy 83:773–783

Ravigururajan TS, Bergles AE (1985) General correlations for pressure drop and heat transfer for single-phase turbulent flow in internally ribbed tubes. In: Augmentation of heat transfer in energy systems, ASME Symp. HTD, vol 52, pp 9–20

Ravigururajan TS, Bergles AE (1996) Development and verification of general correlations for pressure drop and heat transfer in single-phase turbulent flow in enhanced tubes. Exp Therm Fluid Sci 13:55–70

Saha SK, Mallick DN (2005) Heat transfer and pressure drop characteristics of laminar flow in rectangular and square plain ducts and ducts with twisted-tape inserts. Trans ASME 127 (9):966–977

Saravanan A, Senthilkumaar JS, Jaisankar S (2016) Performance assessment in V-trough solar water heater fitted with square and V-cut twisted tape inserts. Appl Therm Eng 102:476–486

Sarma (2003) Heat transfer coefficients with twisted tape inserts

Sethumadhavan R, Raja Rao M (1983) Turbulent flow heat transfer and fluid friction in helical wire coil inserted tubes. Int J Heat Mass Transf 26:1833–1845

Sethumadhavan R, Raja Rao M (1986) Turbulent flow friction and heat transfer characteristics of single- and multi-start spirally enhanced tubes. J Heat Transf 108:55–61

Sheriff N, Gumley P (1966) Heat transfer and friction properties of surfaces with discrete roughness. Int J Heat Mass Transf 9:1297–1320

Vicente PG, Garcia A, Viedma A (2002) Experimental study of mixed convection and pressure drop in helically dimpled tubes for laminar and transition flow. Int J Heat Mass Transf 45:5091–5105

Webb RL (1979) Toward a common understanding of the performance and selection of roughness for forced convection. In: Hartnett JP et al (eds) Studies in heat transfer: a Festschrift for E.R.G. Eckert. Hemisphere, Washington, DC, pp 257–272

Webb RL (1991) Advances in shell side boiling of refrigerants. J Inst Refrig 87:75–86

Webb RL, Eckert ERG (1972) Application of rough surfaces to heat exchanger design. Int J Heat Mass Transf 15:1647–1658

Webb RL, Kim NH (2005) Principles of enhanced heat transfer. Taylor & Francis, New York

Webb RL, Robertson GF (1988) Shell-side evaporators and condensers used in the refrigeration industry. In: Shah RK, Subbarao EC, Mashelkar RA (eds) Heat transfer equipment design. Hemisphere, Washington, DC, pp 559–570

Webb RL, Hui TS, Haman L (1984) Enhanced tubes in electric utility steam condensers. In: Sengupta S, Mussalli YF (eds) Heat transfer in heat rejection systems, Book G00265, HTD, vol 37, pp 17–26

Webb RL, Eckert ERG, Goldstein RJ (1971) Heat transfer and friction in tubes with repeated rib roughness. Int J Heat Mass Transf 14:601–617

Williams F, Pirie MAM, Warburton C (1970) Heat transfer from surfaces roughened by ribs. In: ASME symp volume augmentation of convective heat & mass transfer

Withers JG (1980a) Tube-side heat transfer and pressure drop for tubes having helical internal ridging with turbulent/transitional flow of single-phase fluid. Part 1: single-helix ridging. Heat Transf Eng 2(1):48–58

Withers JG (1980b) Tube-side heat transfer and pressure drop for tubes having helical internal ridging with turbulent/transitional flow of single-phase fluid. Part 2: multiple-helix ridging. Heat Transf Eng 2(2):43–50

Ye QY, Chen WL, Marto PD, Tan YK (1987) Prediction of heat transfer characteristics in spirally fluted tubes by studying the velocity profiles and turbulence structure. In: Marto PD, Tanasawa I (eds) 1987 ASME-JSME thermal engineering joint conference, vol 5. ASME, New York, pp 189–194

Zhang YM, Gu WZ, Han JC (1994) Heat transfer and friction in rectangular channel with ribbed or ribbed-grooved walls. ASME J Heat Transf 116:58–65

Chapter 2
Twisted Tape Insert

There are several variations in twisted tape inserts: continuous full-length twisted tape, short-length twisted tape, regularly spaced twisted tape elements, twisted tape with centre clearance, twisted tape with multiple modules, twisted tape with gradually varying twist-pitches, twisted tape with oblique teeth, etc.

The twist ratio, y of the twisted tape is given by

$$y = \frac{H}{D_h} \quad \text{or} \quad \tan \alpha = \frac{\Pi}{2y},$$

where H is the axial distance for 180° twist.

Twisted tapes are fitted to the tube with some clearance for easy insertion and removal of the tape. Sometimes, however, the twisted tape is snugly fitted to the tube. The tape thickness is finite, it creates resistance to flow passage and increases the average flow velocity. The tape-tube wall clearance gives poor thermal contact between the tape and the tube wall, and the heat transfer from the tape becomes quite small.

Twisted tape reduces the hydraulic diameter and increases the heat transfer coefficient even for zero tape twist. The tangential velocity component increases the flow velocity, and it causes fluid mixing due to the two non-axisymmetric velocity profiles. The shear stress is increased at the wall. The increased velocity near the wall and fluid mixing by the secondary flow increases heat transfer coefficient. The more the swirl intensity, the more is the enhancement. Good thermal contact makes the tape act as a fin to increase heat transfer. Loosely fitting tape, however, cannot cause additional heat transfer. Bergles and Joshi (1983) give a useful account of different types of swirl flow devices for laminar flow.

Thorsen and Landis (1968) studied the effect of centrifugal forces causing fluid mixing from the core region with the fluid near the wall. This causes heat transfer enhancement when the flowing fluid is being heated in its passage through the tube from the upstream end to the downstream end. The colder, high-density core region

© The Author(s), under exclusive license to Springer Nature Switzerland AG 2020
S. K. Saha et al., *Insert Devices and Integral Roughness in Heat Transfer Enhancement*, SpringerBriefs in Applied Sciences and Technology,
https://doi.org/10.1007/978-3-030-20776-2_2

fluid is forced outward to mix with the warm, low-density fluid near the wall. For fluid cooling, the centrifugal force acts to maintain thermal stratification of the fluid.

Smithberg and Landis (1964), Lepina and Bergles (1969), Manglik and Bergles (1992a, b), Hong and Bergles (1976), Date (1974, 1973), Date and Saha (1990), Date and Singham (1972), Manglik (1991), Marner and Bergles (1978, 1985), Patil (2000), Saha et al. (2001), Saha et al. (1989), Shivkumar and Raja Rao (1998), Zhang et al. (1997), Zhuo et al. (1992) and Zimparov (2001, 2002, 2004a, b) inter alia are important literature on the investigation on twisted tapes. More recent literature on twisted tape would be found elsewhere in the text.

Laminar flow, turbulent flow as well as the transition flow has been considered. Nusselt number and friction factor are functions of flow parameters like Reynolds number, Grashoff number, Peclet number, Graetz number and Rayleigh number, the geometrical parameters like hydraulic diameter for non-circular ducts and twist ratio of the tape, and fluid properties like Prandtl number, viscosity corrections for non-isothermal pressure drop, etc. For the correlations, the domain and range of parameters are very important. No comprehensive single correlation is available, which can include entire database available.

Laminar flow is influenced by several factors like the thermal boundary condition, entrance region effects, natural convection at low Reynolds number, fluid property variation across the boundary layer and the duct cross-sectional shape. For simultaneously developing velocity and temperature profiles, the local value of Nusselt number will be different, as compared to a fully developed velocity profile. Laminar flow data are taken with uniform wall heat flux or uniform wall temperature boundary condition, usually with a developed velocity profile entering the heated test section.

For uniform wall heat flux boundary condition, the local heat transfer coefficient is obtained using the known heat flux, and the wall temperature is measured by thermocouples along the tube length. We get the local Nusselt number for fully developed flow from wall thermocouples sufficiently far from the heated inlet. However, it is very difficult to get such value using a moderate test section length. For uniform wall thermal boundary condition, the local heat flux is not known. The usual way is to measure the UA value over the test section length and subtract the shell-side resistance to obtain the average tube-side heat transfer coefficient over the test section length. Again, fluid property variation across the boundary layer is very important for laminar flow. This effect is taken care of by $\left(\frac{\mu_b}{\mu_w}\right)^{0.14}$.

The buoyancy force (natural convection) also affects the Nusselt number. This is determined by Rayleigh number. The natural convection is important for a twisted tape for Gr/Sw^2 is >1, and the ratio takes care of the relative influence of buoyancy force and centrifugal forces. So, for laminar flow data, all the effects like entrance effects, thermal boundary condition, fluid property variation across the boundary layer and buoyancy force are important. Figures 2.1, 2.2 and 2.3 show the laminar flow twisted tape data. It is evident that the performance of the twisted tape is better in heating than in cooling (of course, this will depend on Prandtl number of the

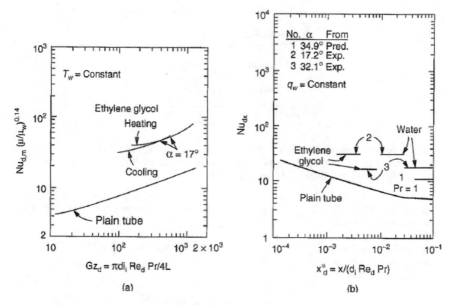

Fig. 2.1 Nusselt number for laminar flow in tubes with twisted tape insert: (**a**) constant wall temperature, (**b**) constant heat flux (from Bergles and Joshi 1983)

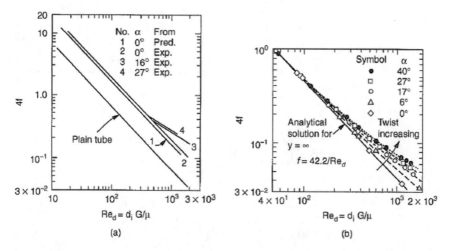

Fig. 2.2 Friction factor in tubes with twisted tape insert for laminar flow: (**a**) experimental and predicted, (**b**) predicted by Date (1974) for zero tape thickness (from Bergles and Joshi 1983)

fluid). Also, the performance of the internally finned tube is better than that of the twisted tape.

Works on predictive methods, data and correlations for laminar flow are available in Date (1974), Date and Singham (1972), Hong and Bergles (1976) and Saha et al. (2001) (Figs. 2.4 and 2.5).

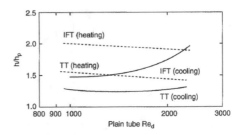

Fig. 2.3 Performance comparison of internally finned tube for laminar flow of ethylene glycol (IFT, 16 fins, $e/D_i = 0.084$, $\alpha = 27°$), and twisted tape insert (TT, $\alpha = 30°$, $y = 5.4$, $t/D = 0.053$) for laminar flow using case FG-2a (Webb and Kim 2005)

Fig. 2.4 Asymptotic behaviour of isothermal friction factor for fully developed laminar flow with twisted tape insert (from Manglik and Bergles 1992a)

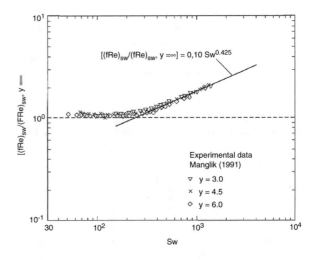

$$\frac{(fRe_d)_{sw}}{(fRe_d)_{sw,\,y=\infty}} = \left(1 + 10^{-6}Sw^{2.55}\right)^{-1/6} \tag{2.1}$$

where

$$(fRe_d)_{sw,\,y=\infty} = 15.767\left(\frac{\pi + 2 - 2t/d_i}{\pi - 4t/d_i}\right) \tag{2.2}$$

$$f_{sw} = \frac{\Delta p d_i}{2\rho u_{sw}^2 L_s} = f\frac{L}{L_s}\left(\frac{u_c}{u_{sw}}\right)^2 \tag{2.3}$$

$$Nu_d = 5.172\left[1 + 0.005484\left(\frac{Re_d}{y}\right)^{1.25}Pr^{0.7}\right]^{0.5} \tag{2.4}$$

Xie et al. (1992) gave turbulent flow correlations.

Fig. 2.5 Effect of
shortening the twist tape
length for oil in laminar flow
($205 \leq Pr \leq 518$) for $y = 5$.
(**a**) Friction
factor vs. Reynolds number,
(**b**) Nusselt
number vs. Reynolds
number (from Saha et al.
2001)

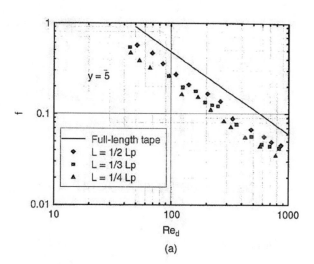

(a)

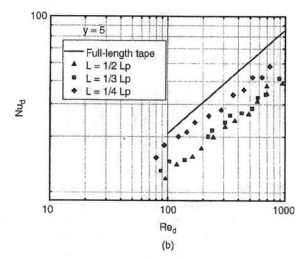

(b)

$$Nu_d = 0.0149Re_{d,f}{}^{0.833}Pr^{2/3}y^{-0.493}\left(\frac{\mu}{\mu_w}\right)^{0.14} \tag{2.5}$$

$$f = \left(3.61 + 8.92y - 1.52y^2\right)Re_{d,f}{}^{-\left(0.28+0.14y-0.021y^2\right)}\left(\mu/\mu_w\right)^{1/3} \tag{2.6}$$

Information on uniform wall temperature may be obtained from Manglik (1991),
Manglik and Bergles (1992a) and Marner and Bergles (1978, 1985).

$$Nu_{d,m} = 4.612\left(1 + 0.0951Gz^{0.894}\right)^{0.5}\left(\mu/\mu_w\right)^{0.14} \tag{2.7}$$

For fully developed flow ($Gz \ll Sw$),

$$Nu_{d,m} = 0.106Sw^{0.767}Pr^{0.3}(\mu/\mu_w)^{0.14} \tag{2.8}$$

As $S_w \to \infty$, $Gz \to 0$,

$$Nu_{d,m} = 4.612\left(1 + 6.413 \times 10^{-9}\left(SwPr^{0.391}\right)^{3.835}\right)^{0.2}(\mu/\mu_w)^{0.14} \tag{2.9}$$

Combining Eqs. 2.8 and 2.10,

$$
\begin{aligned}
Nu_{d,m} \\
= 4.612\left[\left(1 + 0.0951Gz^{0.894}\right)^{2.5} + 6.413 \times 10^{-9}\left(SwPr^{0.391}\right)^{3.835}\right]^{0.2}(\mu/\mu_w)^{0.14}
\end{aligned}
\tag{2.10}
$$

For $Gr >> Sw^2$,

$$\frac{Nu_{d,m}}{4.612} = 4.294 \times 10^{-2}(Re_d Ra)^{0.223} \tag{2.11}$$

The final form of the equation is as follows:

$$
Nu_{d,m} = 4.612\left\langle \begin{array}{c} \left[\left(1 + 0.0951Gz^{0.894}\right)^{2.5} + 6.413 \times 10^{-9}\left(SwPr^{0.391}\right)^{3.835}\right]^{0.2} \\ +2.132 \times 10^{-14}(Re_d Ra)^{0.223} \end{array} \right\rangle
\times (\mu/\mu_w)^{0.14}
\tag{2.12}
$$

The works were with ethylene glycol ($75 < Pr < 180$), polybutene and water ($3 < Pr < 10$). Patil (2000) worked with varying tape width in laminar flow of a pseudoplastic power-law fluid under uniform wall temperature boundary condition. Expectedly, reduced tape width had poor enhancements.

Turbulent flow twisted tape data may be obtained from Smithberg and Landis (1964), Zhuo et al. (1992), Date (1973), Manglik and Bergles (1992b) and Lepina and Bergles (1969) (Fig. 2.6).

$$\frac{f}{f_p} = \left(\frac{y}{y-1}\right)^m \tag{2.13}$$

$$m = 1.15 + \frac{1.25(70,000 - Re_{Dh})}{65,000} \tag{2.14}$$

$$f_p = 0.046Re_d^{-0.2} \tag{2.15}$$

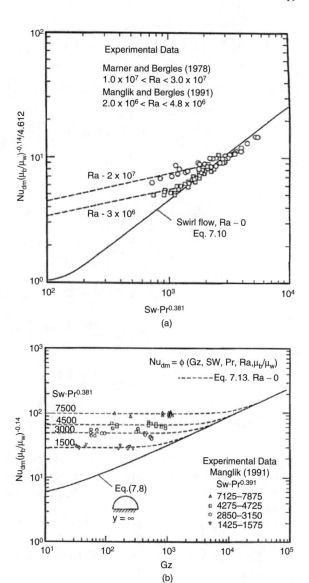

Fig. 2.6 Comparison of predictions using data illustrating the effect of (**a**) natural convection (*Ra*) and (**b**) entrance length (*Gz*) (from Manglik and Bergles 1992a)

$$f = \frac{0.079}{Re_d^{0.25}} \left(\frac{\pi}{\pi - 4t/d_i} \right)^{1.75} \left(\frac{\pi + 2 - 2t/d_i}{\pi - 4t/d_i} \right)^{1.25} \left(1 + \frac{2.752}{y^{1.29}} \right) \qquad (2.16)$$

$$\frac{Gr}{Re_{Dh}^2} = \frac{2D_h \beta_T \Delta T \tan \alpha}{d_i} \qquad (2.17)$$

$$Nu = 0.021F\left(1 + 0.25\sqrt{\frac{Gr}{Re_{\mathrm{Dh}}}}\right)Re_{\mathrm{Dh}}^{0.8}Pr^{0.4}\left(\frac{T_{\mathrm{w}}}{T_{\mathrm{h}}}\right)^{-0.32} \quad \text{for heating} \quad (2.18)$$

$$Nu = 0.021F\left(1 - 0.25\sqrt{\frac{Gr}{Re_{\mathrm{Dh}}}}\right)Re_{\mathrm{Dh}}^{0.8}Pr^{0.3}\left(\frac{T_{\mathrm{w}}}{T_{\mathrm{h}}}\right)^{-0.1} \quad \text{for cooling} \quad (2.19)$$

$$F = 1 + 0.004872\frac{\tan^2\alpha}{d_{\mathrm{i}}(1 + \tan\alpha)} \tag{2.20}$$

$$Nu_{\mathrm{sc}} = \frac{h_{\mathrm{sc}}D_{\mathrm{h}}}{k} = 0.023Re_{\mathrm{Dh}}^{0.8}Pr^{0.4} \tag{2.21}$$

$$Re_{\mathrm{Dh}} = \frac{u_{\mathrm{sw}}D_{\mathrm{h}}}{\nu}, \quad \text{where} \quad u_{\mathrm{sw}} = u_{\mathrm{c}}\sqrt{1 + \tan^2\alpha} \tag{2.22}$$

$$g_{\mathrm{r}} = \frac{2u_{\theta}^2}{d_{\mathrm{i}}} = \frac{2}{d_{\mathrm{i}}}\left(\frac{u_{\mathrm{c}}\pi}{2y}\right)^2 = \frac{4.94}{d_{\mathrm{i}}}\left(\frac{u_{\mathrm{c}}}{y}\right)^2 \tag{2.23}$$

$$Nu_{\mathrm{cc}} = \frac{h_{\mathrm{cc}}D_{\mathrm{h}}}{k} = 0.12(Gr_{\mathrm{Dh}}Pr)^{1/3} \tag{2.24}$$

$$Gr_{\mathrm{Dh}} = \frac{4.94\beta_{\mathrm{T}}\Delta T_{\mathrm{fs}}D_{\mathrm{h}}Re_{\mathrm{Dh}}^2}{d_{\mathrm{i}}y^2} \tag{2.25}$$

$$\frac{Nu_{\mathrm{d}}}{Nu_{\mathrm{d},y=\infty}} = 1 + \frac{0.769}{y} \tag{2.26}$$

$$Nu_{\mathrm{d},y=\infty} = 0.023Re_{\mathrm{d}}^{0.8}Pr^{0.4}\left(\frac{\pi}{\pi - 4t/d_{\mathrm{i}}}\right)^{0.8}\left(\frac{\pi + 2 - 2t/d_{\mathrm{i}}}{\pi - 4t/d_{\mathrm{i}}}\right)^{0.2}\phi \tag{2.27}$$

Gupte and Date (1989) and Coetzee et al. (2001) studied the performance of twisted tapes in annuli. Zimparov (2001, 2002, 2004a, b), Zhang et al. and Shivkumar and Raja Rao (1998) studied the performance of twisted tapes in rough tubes.

Saha et al. (1989, 2001) and Xie et al. (1992) worked with regularly spaced twisted tape elements (Figs. 2.7 and 2.8, Table 2.1), and Reynolds number of the segmented tape is smaller than that of the continuous tape for constant pumping power and is given by

$$\frac{Re_{\mathrm{st}}}{Re_{\mathrm{ct}}} = \left(\frac{f_{\mathrm{st}}A_{\mathrm{c,ct}}}{f_{\mathrm{ct}}A_{\mathrm{c,st}}}\right)1/3 \tag{2.28}$$

Abolarin et al. (2019) studied the effect of alternating clockwise and counter-clockwise twisted tape (CCCTT) inserts on heat transfer and pressure drop characteristics in the transitional flow. The inserts were made of copper plate strips with a length, width and thickness of 450 mm, 18 mm and 1 mm, respectively. Twelve copper strips were twisted to obtain five twist ratio, and clockwise direction twisted

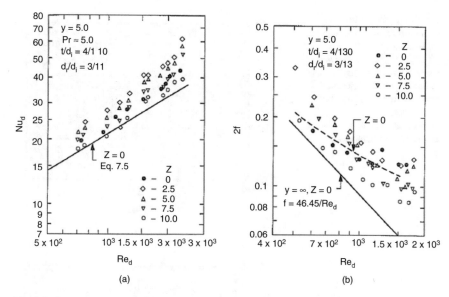

Fig. 2.7 Developing laminar flow of water with q = constant in a tube having a segmental twisted tape of $y = 5$ and $0 < z < 10$: (**a**) Nusselt number, (**b**) friction factor (from Saha et al. 1989)

tape was connected to counter clockwise direction twisted tape. The connection angles of CCCTT inserts was 0°, 30° and 60° as shown in Fig. 2.9. The CCCTT inserts were placed inside the smooth circular copper tube with an inner diameter of 19 mm. The tube was subjected under the constant heat fluxes of 1.35, 2, 3 and 4 kW/m² and water was used as working fluid with Reynolds number range of 300–11,404. They focused mainly in the region of transition flow and observed the effect of connection angle and heat flux on both the start and the end of the transitional flow regime.

They observed that heat transfer can be enhanced by increasing the connection angle in transitional flow regime. They found heat transfer enhanced by an increase in heat flux in the laminar flow regime and delayed in transitional regime. They developed the correlations for the prediction of heat transfer and pressure drop characteristics as a function of Reynolds number, modified Grashof number and connection angle in the laminar, transition and turbulent regimes. Bandyopadhyay et al. (1991), Safikhani and Abbasi (2015), Eiamsa-ard and Promvonge (2010a, b), Bhattacharyya and Saha (2012), Chakroun and Al-Fahed (1996), Meyer and Abolarian (2018), Klepper (1972) and Bhuiya et al. (2013) worked on heat transfer enhancement by using twisted tape inserts and with some modified twisted tapes.

Qi et al. (2018) studied the effect of twisted tape structures on thermo-hydraulic performances of nanofluids in a triangular tube. They investigated experimentally the effect of nanoparticle mass fraction (ω = 0.1 wt%, 0.3 wt% and 0.5 wt%), different twisted tape length (P = 25 mm, 40 mm, 55 mm, 65 mm, 75 mm), Reynolds number (Re = 400–9000) on the heat transfer coefficient and resistance coefficient enhancement ratio. Figure 2.10 shows the detailed specification of

Fig. 2.8 Effect of number of segmented tape sections for laminar oil flow ($205 \leq Pr \leq 518$) with y and $z = 2.5$ for short-length twisted tapes. (**a**) Friction factor vs. Reynolds number, (**b**) Nusselt number vs. Reynolds number (from Saha et al. 2001)

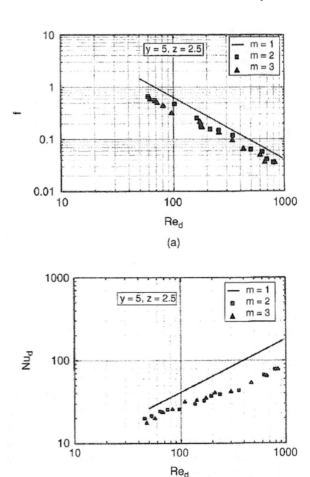

(a)

(b)

Table 2.1 Case FG-1 PEC for segmented twisted tapes (Webb and Kim 2005)

z	Nu_{st}/Nu_{ct} for y values of			
	3.18	5.0	7.5	10.0
2.5	1.44	1.47	1.28	1.08
5.0	1.34	1.31	1.13	1.05
7.5	1.17	1.16	1.12	1.04
10	1.09	1.11	1.09	1.00

triangular tube and twisted tape. It was observed that triangular tube with twisted tape increased the Nusselt number by 52.5% and 34.7% in laminar and turbulent flow, respectively, compared with smooth tube.

It was observed that comprehensive performance index decreases with increase in Reynolds number in laminar flow and then increases with increase in Reynolds

(d)   Counter clockwise $\theta = 0°$ Clockwise

(e)   Counter clockwise   Clockwise

$\theta = 30°$

(f)   Counter clockwise   Clockwise

$\theta = 60°$

Fig. 2.9 Schematic representation of the CCCTT insert with connection angles θ of 0°, 30° and 60° (Abolarin et al. 2019)

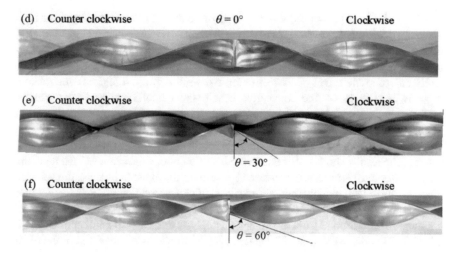

(a)

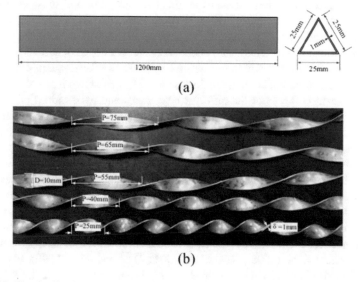

(b)

Fig. 2.10 The details of (**a**) the triangular tube and (**b**) the twisted tape (Qi et al. 2018)

number in turbulent flow. They found from the experimental results that mass fraction of large nanoparticles and small length of each twisted tape unit had high comprehensive performance index. It was also observed that triangular tube with twisted tape has more advantage than corrugated tube in laminar flow. Kumar et al. (2017), Goudarzi and Jamali (2017) and Sundar et al. (2017, 2018) investigated the heat transfer performance of nanofluid flow in a tube with inserts.

Liu et al. (2018) proposed the new type of coaxial cross twisted tapes for high viscous oil to enhance the heat transfer rate in laminar flow. They used three varieties of modified twisted tape, namely coaxial cross twisted tape, coaxial cross double-twisted tape (CCDTT) and coaxial cross triple-twisted tape (CCTTT). They numerically analysed the effect of four different twist ratios (2, 3, 4 and ∞) and coaxial cross twisted tapes on the convective heat transfer coefficient and heat transfer performance for a given pumping power in the range of Reynolds number from 40 to 1050.

Table 2.2 shows the various correlations for different types of working fluids and inserts. The results showed that due to the insertion of coaxial cross twisted tape, the heat transfer rate significantly increased because inserts assist to increase the velocity near the tube wall which thus reduces the thickness of boundary layer. Higher temperature gradient formed near the wall increases the convective heat transfer coefficient. It was found from the numerical analysis that Nusselt number of CCDTT and CCTTT tubes were 1.68–4.26 and 3.29–6.68 times higher than that of the plain tube.

Meyer and Abolarin (2018) investigated heat transfer and pressure drop characteristics in the transition zone for a smooth circular tube with twisted tape inserts and a square-edged inlet. They studied the effect of heat flux and twist ratio on the friction factor and flow characteristics. Experiments were carried out in a circular tube with a length of 5.27 m, internal diameter of 19 mm and twist ratios of 3, 4 and 5. The experiments were conducted under constant heat fluxes of 2, 3 and 4 Kw/m^2. Experimental tests were carried out in the Reynolds number range of 400–11,400, and Prandtl numbers varied between 2.9 and 6.7. They used two types of methods to find the transition points of the different heat fluxes and twist ratio: first, standard deviation of the temperature measurement and, second, three linear curve fits on a log-log scale. Table 2.3 shows the geometric values of the experimental equipment.

Correlations were developed for the non-dimensionalized heat transfer coefficient and friction factors. It was observed that Colburn j-factor increased as the twist ratio decreased for the same heat flux. It was found that higher heat fluxes delayed the transition from laminar flow to transitional flow at constant twist ratio. From the experimental results, it was seen that friction factor was varied and inversely proportional to twist ratio. They found that friction factor can be decreased by increasing heat flux, when both the twist ratio and the Reynolds number were kept constant. Zhang et al. (2008), Saha et al. (2001), Date and Saha (1990), Eiamsa-ard (2010), Maddah et al. (2014), Salman et al. (2014) and Sivakumar et al. (2015) also worked on heat transfer enhancement in a laminar flow or turbulent flow or both by using twisted tape inserts in a tube.

Boonloi and Jedsadaratanachai (2016) reported an assessment on heat transfer, pressure loss and thermal performance in a circular tube heat exchanger with modified twisted tapes. They investigated the influences of the hole sizes (LR = 0.3, 0.44, 0.78) and twisted ratios (y/D = 1, 1.5, 2 and 4) on thermo-hydraulic characteristics with numerical method in turbulent regime (Re = 3000–10,000). The rectangular holes were cut out from general twisted tape.

Table 2.2 Summarization of the insertion for the high and low viscosity fluid (Liu et al. 2018)

Author	Working fluid	Parameter	Type	Correlation
Manglik and Bergles (1993a, b)	Water, ethylene glycol	Twist ratio: 3.0, 4.5, 6.0	Twisted tape	$f\left(\left(\left(\frac{15.767}{Re}(1+10^{-6}Sw^{2.55})^{1/6}\left(1+\left(\frac{\pi}{2y}\right)^2\right)\left(\frac{\pi+2-2\delta/d}{\pi-4\delta/d}\right)^2\left(\frac{\pi}{\pi-4\delta/d}\right)\right)^{10}\right.\right.$ $\left.\left.+\left(\frac{0.0791}{Re^{0.25}}\left(\frac{\pi}{\pi-4\delta/d}\right)^{1.75}\left(\frac{\pi+2-2\delta/d}{\pi-4\delta/d}\right)^{1.25}\left(1+\frac{2.752}{y^{1.29}}\right)\right)^{10}\right)\right)^{0.1}$ $Nu=0.023\left(1+\frac{0.769}{y}\right)Re^{0.8}Pr^{0.4}\left(\frac{\pi}{\pi-4\delta/d}\right)^{0.8}\left(\frac{\pi+2-2\delta/d}{\pi-4\delta/d}\right)^{0.2}\Phi$
Naphon [13]	Water	Re: 7000–23,000 Twist ratio: 3.1–5.5	Twisted tape	$Nu=0.648Re^{0.36}\left[1+\frac{D}{H}\right]^{2.475}Pr^{1/3}$ $f=3.517Re^{-0.414}\left[1+\frac{D}{H}\right]^{10.45}$
Murugesan et al. [14]	Water	Re: 2000–12,000 Twist ratio: 2.0, 4.4, 6.0	Twisted tape consisting of wire nails	$Nu=0.063Re^{0.789}y^{-0.257}Pr^{0.33}$ $f=28.91Re^{-0.731}y^{-0.255}$
Sivashanmugam et al. [15]	Water	Twist ratio: 1.95, 2.93, 3.91, 4.89	Helical screw tape	$Nu=0.017Re^{0.996}y^{-0.5437}Pr$ $f=10.7564Re^{-0.387}y^{-1.054}$
Eiamsa-ard et al. [16]	Water	Re: 3000–10,000 Twist ratios: 2.5, 5.0	Loose fit twisted tapes	The thermal performance factor is affected by tape clearance ratios, and tight fit twisted tapes gave best thermal performance factor
Zhang et al. [17]	Water	Re: 300–1800 Clearance ratio: 0.25, 0.3 and 0.35	Multiple regularly spaced twisted tapes	$Nu=2.0358Re^{0.238}a^{0.0492}$ $f=111.2919Re^{-0.7236}a^{0.6071}$ (for triple-twisted tape) $Nu=1.5689Re^{0.2629}a^{-0.0773}$ $f=181.5216Re^{-0.732}a^{0.7234}$ (for quadruple-twisted tape)

(continued)

Table 2.2 (continued)

Author	Working fluid	Parameter	Type	Correlation
Agarwal and Rao [18]	Servotherm oil	Twist ratios: 2.41–4.84 Re: 70–4000 Pr: 195–375	Twisted tapes	$Nu_a = 0.725 Re_s^{0.568} y^{-0.788} Pr^{1/3}\left(\frac{\mu_b}{\mu_w}\right)^{0.14}$ (heating) $Nu_a = 1.365 Re_s^{0.517} y^{-1.05} Pr^{1/3}\left(\frac{\mu_b}{\mu_w}\right)^{0.14}$ (cooling) $(f_a Re_s y^{0.28}) = \left[\left\{75.74\left(\frac{Re_s}{y}\right)^{0.0216}\right\}^{10} + \left\{19.48\left(\frac{Re_s}{y}\right)^{0.3481}\right\}^{10}\right]^{0.10}$
Saha [20]	Servotherm medium oil	Corrugation pitch: 2.0437, 5.6481 Corrugation angle: 30°, 60° Centre clearance: 0, 0.2, 0.4, 0.6	Circular tube having axial corrugations and fitted with centre-cleared twisted tape	$Nu_m = \left[\left[(1+0.0798.5Gz^{0.9335})^{.25} + 8.2365 \times 10^{-6}(SW Pr^{0.565})^{2.655}\right]^{2.0}\right.$ $\left. + 1.5447 \times 10^{-15}(Re_a Ra)^{2.18}\right]^{0.1}\left(\frac{\mu_b}{\mu_w}\right)^{0.14}\left(1+\frac{(1+\exp(0.093c))\exp(0.0839\sin\theta)}{(p/e)^{0.605}}\right)$ $(fRe)_{sw} = 17.355\left(\frac{\pi+2-2\delta/D_h}{\pi-4\delta/D_h}\right)^2 (1+10^{-6}SW^{2.67})^{1/7}\left(1+\frac{(1+\exp(0.088c))\exp(0.0421\sin\theta)}{(p/e)^{0.614}}\right)$
Saha and Saha [21]	Servotherm medium oil	Screw tape parameter: ∞, 0.41, 031, 0.25 Rib height: 0.07692, 0.1026 Rib helix angle: 30°, 60°	Circular tube having integral helical rib roughness with helical screw tapes	$Nu_m = 5.172\left[\left[(1+0.06938Gz^{0.9252})^{2.5} + 8.1755 \times 10^{-6}(SW Pr^{0.565})^{2.655}\right]^{2.0}\right.$ $\left. + 1.5231 \times 10^{-15}(Re_a Ra)^{2.18}\right]^{0.1}\left(\frac{\mu_b}{\mu_w}\right)^{0.14}$ $\left(1+\frac{(1+\exp(0.08857p))\exp(0.0113\sin\alpha)}{(e/D_h)^{0.618}}\right)$ $(fRe)_{sw} = 17.355\left(\frac{\pi+2-2\delta/D_h}{\pi-4\delta/D_h}\right)^2 (1+10^{-6}SW^{2.67})^{1/7}\left(1+\frac{(1+\exp(0.0812p))\exp(0.05632\sin\alpha)}{(e/D_h)^{0.626}}\right)$

| Rout and Saha [23] | Servotherm medium oil | Screw tape parameter: 1, 0.41, 031, 0.25 Wire coil helix angle: 30°, 60° Coil wire diameter: 0.07692, 0.1026 | Wire coil and helical screw tape | $$Nu_m = 5.172 \left[\left(1 + 0.06958 Gz^{0.8825} \right)^{2.5} + 8.1332 \times 10^{-6} \left(SW Pr^{0.565} \right)^{2.655} \right]^{2.0}$$ $$+ 1.5638 \times 10^{-15} \left(Re_a Ra \right)^{2.18} \Big]^{0.1} \left(\frac{\mu_b}{\mu_w} \right)^{0.14} \left(1 + \frac{(1 + \exp(0.08858 p)) \exp(0.0813 \sin\theta)}{(d/D_h)^{-0.6682}} \right)$$ $$f = 17.355 \left(\frac{\pi + 2 - 2\delta/D_h}{\pi - 4\delta/D_h} \right)^2 \left(1 + 10^{-6} SW^{2.67} \right)^{1/7} \left(1 + \frac{(1 + \exp(0.0785 p)) \exp(0.05538 \sin\theta)}{(d/D_h)^{-0.693}} \right)$$ |

Table 2.3 Parametric values of the experimental setup (Meyer and Abolarin 2018)

Parameters	Symbols	Values
Thickness of twisted tape inserts	δ	1 mm
Outer diameter of test section	D_e	22.0 mm
Inner diameter of test section	D_i	19.0 mm
Pitches of the twisted tape inserts	H	54. 72, 90 mm
Length of twisted tape/test section	L	5.27 m
Heat transfer length	L_h	4.8 m
Pressure drop length	$L_{\Delta p}$	2.45 m
Prandtl numbers	Pr	2.9–6.7
Heat fluxes	$\dot{q}$	2, 3, 4 kW/m^2
Reynolds numbers	Re	400–11,400
Inlet temperature of the working fluid	T_i	20 °C
Width of twisted tape insert	W	18 mm
Twist ratios	y	3, 4, 5

Numerical results were compared for smooth tube and regular twisted tape. It was found that modified twisted tape created longitudinal vortex flows that helped to increase fluid mixing, thus heat transfer rate became higher than that of smooth tube. It was found that holes of twisted tape helped to reduce the pressure loss. The numerical results stated that maximum thermal enhancement factor was around 1.31 and 1.39 for the single-twisted tape and double-twisted tape, respectively, at $R = 3000$, LR $= 0.78$ and TR $= 1$. Piriyarungrod et al. (2015), Promvonge (2015), Hindasageri et al. (2015), Rios-Iribe et al. (2015), Chokphoemphun et al. (2015) and Kanizawa et al. (2014) also evaluated the heat transfer enhancement and pressure drop penalty of the working fluid inside the tube with twisted tape inserts.

Hong et al. (2017) experimentally investigated the turbulent thermo-hydraulic characteristics in a plain tube by using overlapped multiple twisted tapes (MTTs). The value of overlapped twist ratios varied from 0.74 to 2.97, Reynolds number from 5800 to 19,200 and tape number changed from 3 to 5. They used air as working fluid. Heat transfer tests and pressure drop experiments were conducted at constant heat flux and isothermal conditions, respectively. They compared the values of Nusselt number, friction factor, overall thermal performance evaluation criterion (PEC), entropy generation and entransy dissipation of MTTs.

It was found from the experimental results that Nusselt number and friction factor increases with increasing tape number and decreasing overlapped twist ratio. They observed that entropy generation can be reduced by increasing the tape number and decreasing the overlapped twist ratio. They developed thermo-hydraulic empirical correlations with deviations of $\pm 5\%$ and compared the experimental results with previous studies. Eiamsa-ard et al. (2013, 2015a, b), Chokphoemphun et al. (2015), Hong et al. (2012), Promvonge (2015), Tamna et al. (2016), Chamoli et al. (2017), Jaisankar et al. (2009a, b), Holman (2009) and Bas and Ozceyhan (2012) investigated the heat transfer enhancement in a tube with modified inserts.

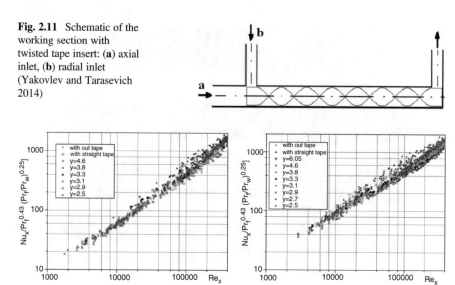

Fig. 2.11 Schematic of the working section with twisted tape insert: (**a**) axial inlet, (**b**) radial inlet (Yakovlev and Tarasevich 2014)

Fig. 2.12 Heat transfer in tubes with (**a**) axial inlet and (**b**) radial inlet (Yakovlev and Tarasevich 2014)

Yakolev and Tarasevich (2014) experimentally studied the heat transfer augmentation performance of twisted tape in tube flow. The tube with axial and radial inlets was considered. They addressed the need for studying the effect of inlet and outlet conditions of the flow on heat transfer characteristics. Distilled water was considered as the working fluid. The entry of coolant has been shown in Fig. 2.11. The heat transfer in tubes with axial inlet and radial inlet has been shown in Fig. 2.12a, b, respectively.

Seemawute and Eiamsa-ard (2010) examined the flow characteristics associated with the twisted tape. The tape having alternate-axis has been described in Fig. 2.13. The objective of the study was to find the best fit between typical twisted tape and alternate-axis twisted tape. They observed better mixing and enhanced heat transfer using alternate-axis tape than that by using typical twisted tape. They visualized the effect via a dye injection technique and found that swirl number and residence time increased with decrease in twist ratio.

Chang and Guo (2012) investigated the pressure drop and thermal characteristics of the tubular flow with enhanced smooth and spiky twisted tapes. The spiky twisted tape performance intensified with the perforated jagged and notch winglets. They compared five modified twisted tapes, namely (1) perforated twist tape (PT), (2) perforated twist tape with jaggedness (PJT), (3) perforated spiky twist tape (PST), (4) perforated spiky twist tape with jaggedness (PJST) and (5) V-notched spiky twist tape (VST) with twist ratios 1.875, 2.186 and 2.815 for each and every tape. This has been represented in Fig. 2.14.

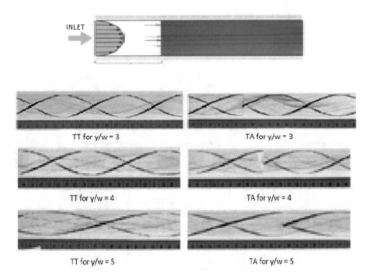

Fig. 2.13 Axial and swirl flow patterns by dye injection techniques (Seemawute and Eiamsa-ard 2010)

Also, they presented Table 2.4 in which various types of modified twisted tapes are summarized with f/f_∞, Nu/Nu_∞ and thermal performance parameter (TPF). They considered both laminar $Re < 2000$ and turbulent regime $Re > 3000$. The modified tapes were 1.5 mm thick, 15 mm wide (w) and 300 mm long (L) made up of stainless straight tape. They experimentally evaluated the local Nusselt number by

$$Nu = {}^{q_f}d/_{\{(T_w - T_f)\kappa_f\}} \tag{2.29}$$

where q_f is local convective heat flux, T_w wall temperature at inner bore, T_f is local fluid bulk temperature and K_f is thermal conductivity of fluid. They measured the axial distribution of Nu along the tube at $Re = 2000$, 10,000, 20,000 and 40,000 for each and every modified twisted tape. This is presented in Fig. 2.15. The Fanning friction factor was calculated from

$$f = \left(\frac{\Delta P}{0.5\rho W_m^2}\right)\left(\frac{d}{4L}\right) \tag{2.30}$$

and thermal performance factor TPF was calculated as

$$\text{TPF} = \left(\frac{\bar{Nu}}{Nu_\infty}\right)\left(\frac{f}{f_\infty}\right)^{1/3} \tag{2.31}$$

However, the Fanning friction factor exponentially decays according to the correlation

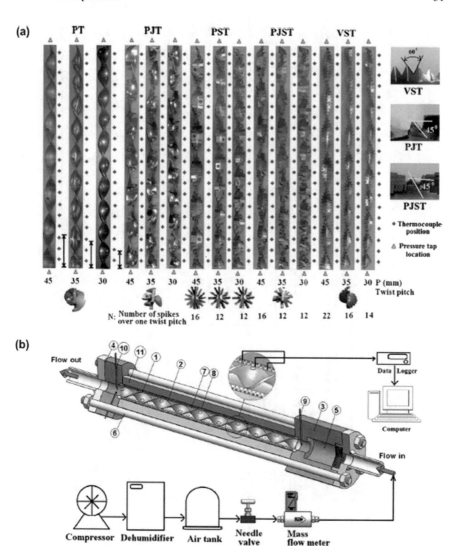

Fig. 2.14 (**a**) Five types of PT, PJT, PST, PJST and VST twisted tapes with $y = 1.875$, 2.186 and 2.815 and (**b**) test module (Chang and Guo 2012)

$$f = c_o + c_1 e^{-c_2 Re} \qquad (2.32)$$

where $c_o - c_2$ are functions of twist ratio. This is presented in Table 2.5. The E, K, M coefficients for Fanning friction factor are presented in Table 2.6. They observed that all modified twisted tapes had the same result trend. The f/f_∞ ratio increased with increase in Reynolds number or decrease in twist ratio (y). The variations of f/f_∞ plotted against Re for each of the tapes are presented in Fig. 2.16, which shows that spiky twisted tape-filled tubes (f/f_∞, range 9.93–58.7) are significantly better than

Table 2.4 The values of f/f_1, Nu/Nu_1 and TPF of modified twist tapes at laminar ($Re < 2000$) and turbulent ($Re > 3000$) reference conditions (Chang and Guo 2012)

Schematic of twisted tape	Re range	f/f_∞		Nu/Nu_∞		TPF	
		$Re < 2000$	$Re > 3000$	$Re < 2000$	$Re > 3000$	$Re < 2000$	$Re > 3000$
Multiple twisted tape							
Twin tapes / Triple tapes y/W: 1.67	1500–14,000	Twin twisted tapes					
		20.74–24.68	15.76–12.61	9.77–12.10	2.45–1.73	3.56–4.15	0.97–0.74
		Triple-twisted tapes					
		25.88–30.44	19.21–14.36	11.43–14.31	3.03–2.58	3.87–4.58	1.13–1.06
(Co-swirl) Twin tapes y/W: 2.5,3.0,3.5,4	3700–21,000	26.6–11.2		2.07–1.22		1.1–0.92	
(Counter-swirl) Twin tapes y/W: 2.5,3,0.3.5,4	3700–21,000	36.3–13.9		2.85–1.46		1.4–1.01	
Alternate twisted tape							
counter-clockwise / clockwise y/W: 3, 4, 5 0=30°, 60°, 90°	3000–27,000	$830 \le Re \le 1990, 0 = 90°$					
		8.1–6.9		13.5–2.06		1.4–1.25	
		$3000 \le Re \le 27,000$					
		6.66–2.31		2.19–1.18		1.3–1.02	
Notched twisted tape							
y/W: 2.94	2950–11,800	7.17–4.62		1.86–1.52		0.96–0.90	

	Re	Re < 2000	Re > 3000	Re < 2000	Re > 3000	Re < 2000	Re > 3000
y/W : 3	1000–20,000	6.5–18	12.5–3.7	4.2–13	3.3–1.35	2.7–4.8	1.44–0.87
y/W : 2.2, 4.4, 6	2000–12,000	5.2–2.5		1.92–1.49		1.27–1.09	
y/W : 4	4000–20,000	3.5–2.2		1.86–1.48		1.175–0.965	
y/W : 3	5000–20,000	(PT) 26.9–14.2 (PT-A)		1.08–0.86		1.13–0.87	
		49.6–24.5		2.8–1.7		1.25–0.85	
y/W : 2, 4.4, 6	3000–11,000	26.7–10.3		3.17–1.79		1.18–1.02	
Jagged and winglet twisted tapes							
y/W : 2.94	2950–11,800	8.7–6.51		2.49–1.96		1.21–1.05	
Straight delta-winglet twisted taped (S-DWT) (jagged twisted tape)	3000–27,000	35.3–10.8		2.5–1.3		1.22–0.85	
Oblique delta-winglet twisted taped (O-DWT) (jagged twisted tape)							
y/W : 3, 4, 5	3000–27,000	44–16.6		2.4–12		1.24–0.86	

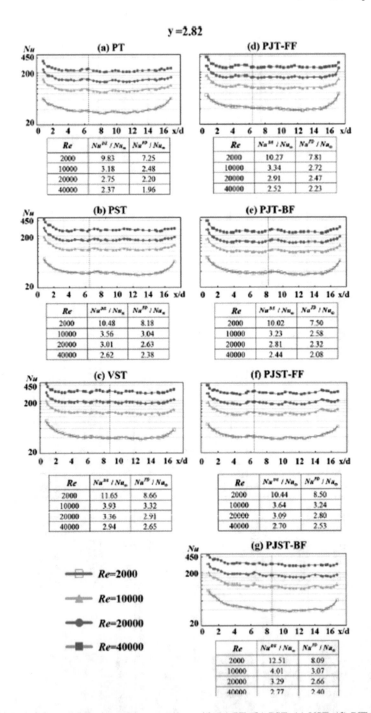

Fig. 2.15 Axial *Nu* distributions along test tubes with (**a**) PT, (**b**) PST, (**c**) VST, (**d**) PJT-FF, (**e**) PJT-BF, (**f**) PJST-FF and (**g**) PJST-BF of $y = 2.82$ at $Re = 2000$, 10,000, 20,000, 40,000 (Chang and Guo 2012)

Table 2.5 c_0–c_2 (c_3) coefficients in f correlation (Chang and Guo 2012)

	$y = 1.88$			$y = 2.19$			$y = 2.82$		
	c_0	c_1	$c_2 \times 1000$	c_0	c_1	$c_2 \times 1000$	c_0	c_1	$c_2 \times 1000$
PT	0.0879	1.14	0.756	0.0769	0.946	0.654	0.0686	0.581	0.6
PJT-FF	0.142	0.797	0.68	0.121	0.638	0.585	0.096	0.48	0.521
PJT-BF	0.145	0.468	0.42	0.107	0.415	0.408	0.094	0.386	0.38
PST	0.152	1.13	0.635	0.13	0.76	0.55	0.115	0.712	0.497
PJST-FF	0.346	0.309	0.54	0.216	0.283	0.508	0.164	0.273	0.43
PJST-BF	0.345	0.679	0.95	0.202	0.611	0.5	0.14	0.519	0.4

VST	$y = 1.88$		$y = 2.19$		$y = 2.82$	
	c_0	c_1	c_0	c_1	c_0	c_1
	0.254	0.348	0.21	0.238	0.17	0.237
	$c_2 \times 1000$	$c_3 \times 100{,}000$	$c_2 \times 1000$	$c_3 \times 100{,}000$	$c_2 \times 1000$	$c_3 \times 100{,}000$
	0.6	0.147	0.55	0.032	0.42	0.018

Table 2.6 E, K, M coefficients in c_0–c_2 (c_3) functions (Chang and Guo 2012)

	$c_{0(y)}$			$c_{1(y)}$			$c_{2(y)}$		
	E	K	M	E	K	M	$E \times 1000$	$K \times 1000$	M
PT	0.0174	0.13	0.339	0.02	4.41	0.725	0.34	1.05	0.512
PJT-FF	0.0174	0.311	0.493	0.02	2.21	0.565	0.34	1.22	0.699
PJT-BF	0.0174	0.35	0.567	0.02	0.648	0.209	0.34	0.314	0.719
PST	0.0174	0.249	0.34	0.02	2.77	0.527	0.34	1.06	0.702
PJST-FF	0.0174	1.89	0.96	0.02	0.364	0.134	0.34	1.09	0.873
PJST-BF	0.0174	3.01	1.21	0.02	1.14	0.295	0.34	177	3.04

VST	c_0		c_1		c_2		c_3	
	E	K	E	K	$E \times 1000$	$K \times 1000$	E	K
	0.0174	0.561	0.02	0.706	0.34	2.33	0	0.00878
	M		M		M		M	
	0.471		0.453		1.13		4.64	

tubes with conventional twisted tapes (f/f_∞, range 3.83–13.41). Similarly, Fig. 2.17 presents the trends of TPF with Reynolds number for each case. The TPF data from Spiky-type tape gradually diminishes with twist ratio due to rise in pressure drop. They concluded that V-notched spiky twist tape (VTS) performance was best among the group challengers, and they proposed that it was the best optimized design for retrofit applications.

Thianpong et al. (2012) experimentally investigated on perforated twisted tapes with parallel wings (PTT) inserted in heating tube. They studied the heat transfer and pressure drop characteristics considering the Reynolds number range 5500–20,500. In comparison to typical twisted tape, modified twisted tape with the new geometries provides better fluid mixing at the expense of increased friction. The flow pattern through the tube with and without the twisted tapes is presented in Fig. 2.18. Here, the fluid flow pattern in the plain tube, twisted tape and perforated twisted tapes were presented via dye visualization technique. They studied the effect of different perforated diameter ratio ($d/W = 0.11$, 0.33 and 0.55) and presented the results in Fig. 2.19. It was observed that decrease in diameter ratio increased the Nusselt number. They concluded that both heat transfer and friction factor were increased with the wing depth ratio (w/W) and the decrease in perforation hole diameter ratio (d/W). They concluded that heat transfer increment dominated the pressure drop loss.

Noothong et al. (2006) studied the consequences of inserting twisted tapes in tube and evaluated pressure drop and heat transfer characteristics. Although many researchers Ray and Date (2003), Kumar and Prasad (2000), Sukhatme et al. (1987), Duplessis and Kroeger (1983) and Lepina and Bergles (1969) worked on heat transfer enhancement, still there is some space for improvement. Noothong et al. (2006) arranged concentric double-pipe heat exchanger made up of Plexiglas and fitted it with twisted tape. This setup has been presented in Fig. 2.20, and the used twisted tape has been presented in Fig. 2.21.

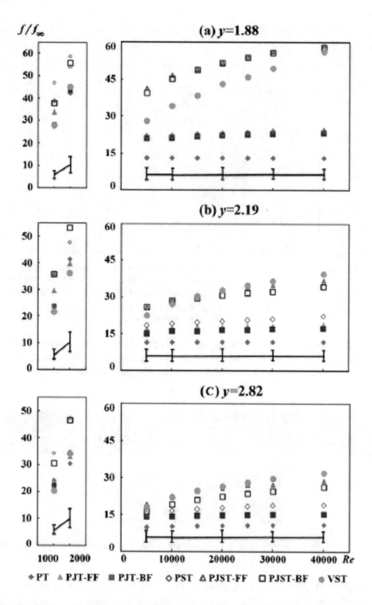

Fig. 2.16 Variations of f/f_1 against Re for PT, PJT-FF, PJT-BF, PST, PJST-FF, PJST-BF and VST tubes at $y =$ (**a**) 1.88, (**b**) 2.18, (**c**) 2.82 and conventional swirl tube with smooth twisted tape (Chang and Guo 2012)

Experimentally, they investigated that there was 188% increase in Nusselt number for twist ratios $y = 5.0$ and 159% for $y = 7.0$ in comparison to plain tube. They found that twisted tape generated swirl and pressure gradient in radial direction, and it encouraged the flow to be turbulent, thus enhancing the convective heat transfer.

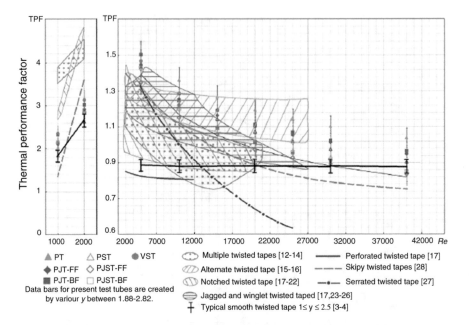

Fig. 2.17 Comparison of TPF variations against *Re* for PT, PJT-FF, PJT-BF, PST, PJST-FF, PJST-BF and VST tubes in *y* range of 1.88–2.82 and conventional swirl tube with smooth twisted tape with the collective TPF zones for the modified twist tapes (Chang and Guo 2012)

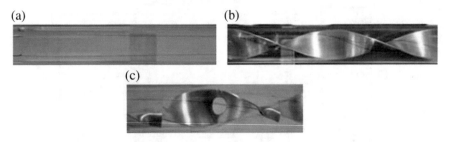

Fig. 2.18 Visualization of flows through (**a**) plain tube, (**b**) TT and (**c**) PTT (Thianpong et al. 2012)

They observed that for low Reynolds number, the increase in Nusselt number was low and vice versa. Also, they concluded that twisted tape increased the friction factor by 3.37 times for *y* = 5.0 and 2.95 times for *y* = 7.0 in comparison to plain tube. The results can be summarized as enhancement efficiency, and Nusselt number raised by decreasing the twisted tape twist ratio.

Sarada et al. (2010) experimentally investigated the heat transfer enhancement in turbulent regime. They arranged twisted tape inserts in a horizontal tube and changed the width of the tape. They observed that when the width varied from 10 mm to 22 m, the rate of heat transfer raised from 36 to 48%. This enhancement in heat

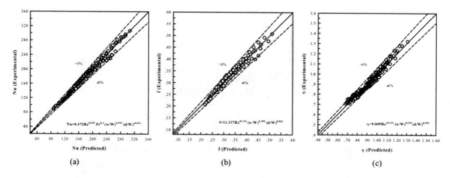

Fig. 2.19 Comparison between the predicted and experimental data of (**a**) *Nu*, (**b**) friction factor and (**c**) thermal performance factor (Thianpong et al. 2012)

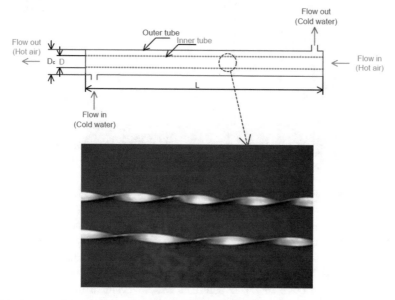

Fig. 2.20 The inner tube fitted with twisted tape at different twist ratios ($y = 5.0$ and 7.0) (Noothong et al. 2006)

transfer is due to centrifugal forces generated which causes spiral motion of fluid and ultimately increases convective heat transfer rate.

Jaisankar et al. (2009a, b) experimentally investigated the thermo-hydraulic performance of thermosyphon solar water heater. They used full-length left–right twist, twist fitted with rod and spacer. They experimentally concluded that the twist fitted with rod was much better than that of spacer.

Eiamsa-ard et al. (2010a, b, c, d, e) investigated the thermo-hydraulic performance of modified twisted tape in a tube. The modified twisted tape was manufactured from a plain tube. The modified twisted tape with centre wings was

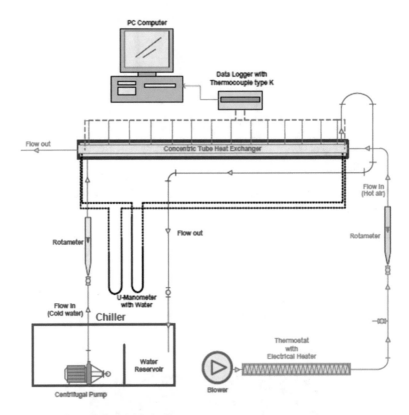

Fig. 2.21 Experimental rig schematic diagram

developed by punching delta-wing with central alignment. Twisted tape consisting of alternate axis was generated by cutting both sides, and lastly twisted tape with centre wings and alternate axes were made by performing operations for centre wing and then moved to operation for alternate wings. These are represented in the Fig. 2.22. The objective of the experiment was to evaluate the effects of these three modifications and find the best design for heat transfer enhancement. Their experimental results were oriented around Nusselt number (Nu), friction factor (f) and thermal performance factor (η) variation with the modified twisted tapes.

They plotted the effect of Nusselt number for modified twisted tape with different angles of attack and compared it with plain twisted tape in Fig. 2.23. Similarly, they plotted friction factor and performance factor in Figs. 2.24 and 2.25, respectively. The flow phenomena associated with the flow having various twisted tapes in the tube are represented in Fig. 2.26. It clearly elaborates the concepts regarding flow pattern in tubes with the above-mentioned inserts. The proposed correlations which govern the friction factor and Nusselt number have been listed in Table 2.7.

Eiamsa-ard et al. (2010a, b, c, d, e) concluded that the combination of wing and alternate axis performed better than that of individual and plain tubes. They found that the pressure drop increased with angle of attack, and it was found to be the

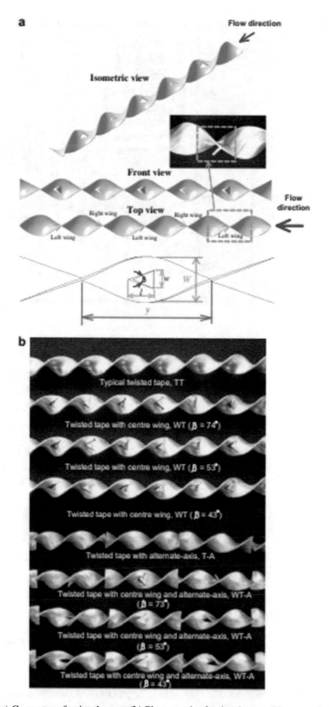

Fig. 2.22 (**a**) Geometry of twisted tapes. (**b**) Photograph of twisted tapes (Eiamsa-ard et al. 2010a, b, c, d, e)

Fig. 2.23 Effect of twisted
tape on Nusselt number: *Nu*
and Nu_t/Nu_p (Eiamsa-ard
et al. 2010a, b, c, d, e)

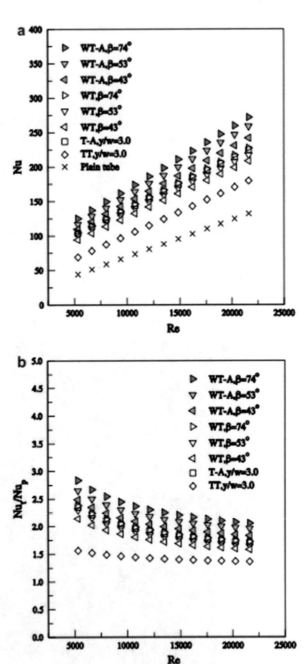

maximum for wing and alternate axis. Also, they observed that the range of Nusselt
number ratio (Nu_t/Nu_p) for wing-embossed twisted tape and wing-embossed twisted
tape with alternate axis was 1.6–2.4 and 1.8–2.8, respectively. However, friction

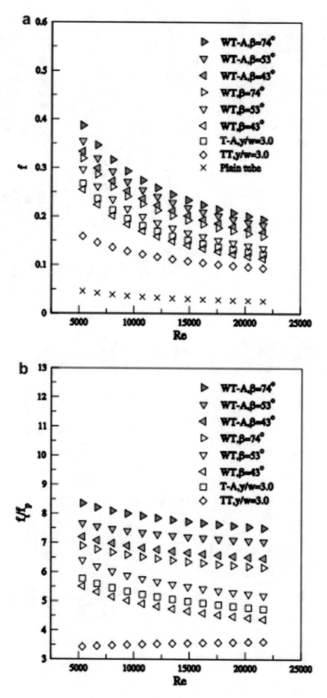

Fig. 2.24 Effect of twisted tape on friction factor (f) and (f_t/f_p) (Eiamsa-ard et al. 2010a, b, c, d, e)

Fig. 2.25 Effect of twisted
tape on thermal performance
factor (Eiamsa-ard et al.
2010a, b, c, d, e)

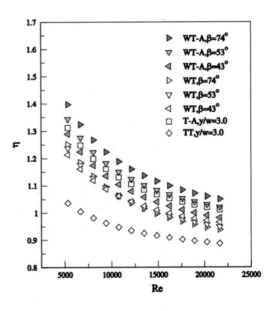

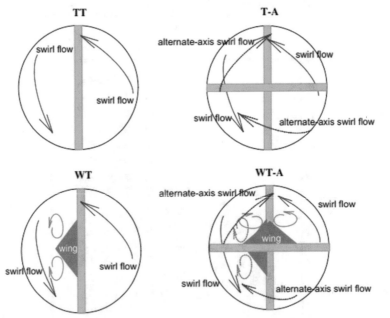

Fig. 2.26 Sketch of flow phenomena in the front view of the tube with various twisted tapes
(Eiamsa-ard et al. 2010a, b, c, d, e)

Table 2.7 Correlations of Nusselt number, friction factor and thermal performance factor (Eiamsa-ard et al. 2010a, b, c, d, e)

Nusselt number	
Plain tube	$Nu = 0.025Re^{0.791} Pr^{0.4}$
Twisted tape with centre wings	$Nu = 0.232Re^{0.595} Pr^{0.4} (1 + \tan \beta)^{0.202}$
Twisted tape with centre wings and alternate axes	$Nu = 0.385Re^{0.568} Pr^{0.4} (1 + \tan \beta)^{0.129}$
Friction factor	
Plain tube	$f = 1.645Re^{-0.416}$
Twisted tape with centre wings	$f = 14.039Re^{-0.505} (1 + \tan \beta)^{0.406}$
Twisted tape with centre wings and alternate axes	$f = 20.445Re^{-0.504} (1 + \tan \beta)^{0.283}$
Thermal performance factor	
Twisted tape with centre wings	$\eta = 4.629Re^{-0.166} (1 + \tan \beta)^{0.067}$
Twisted tape with centre wings and alternate axes	$\eta = 6.772Re^{-0.194} (1 + \tan \beta)^{0.035}$

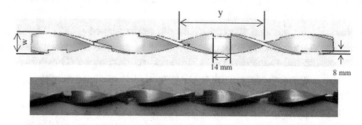

Fig. 2.27 Twisted tape insert (Salam et al. 2013)

factor (f_t/f_p) was 4.4–6.9 and 6.48–8.4 for the wing and wing-embossed transverse axis twisted tape, respectively. Finally, they summarized that wing-embossed transverse axis twisted tape with the highest attack angle ($\beta = 74°$) generated the highest Nusselt number (Nu), friction factor (f) as well as thermal performance factor (η).

Wongcharee and Eiamsa-ard (2011) examined the thermo-hydraulic characteristic performance of clockwise and counterclockwise alternate axis twisted tape. It was the advancement of the work Eiamsa-ard et al. (2010a, b, c, d, e) where they used twisted with transverse axis. Wongcharee and Eiamsa-ard (2011) found that friction factor was higher for twisted tape with alternate axis as compared to plain tube and that it further increased with decrease in twist ratio. It was 50%, 49% and 33% higher than those of twisted tape of twist ratio $y = 3.0$, 4.0 and 5.0, respectively. The Nusselt number was 70.9 and 104.0% higher than that of plain twisted tape having the same twist ratio.

Salam et al. (2013) experimentally investigated the heat transfer characteristics in a circular tube consisting of rectangular-cut twisted tape insert. This rectangular-cut shaped twisted tape was made up of stainless steel and is presented in Fig. 2.27. They experimented in the Reynolds number range of 10,000–19,000 and obtained 68% increased heat transfer as compared to smooth tube. But, 39–80% pressure loss occurred.

Lekurwale et al. (2014) worked on the enhancement of heat transfer by using three types of twisted tape inserts in the turbulent flow condition. The clockwise, counterclockwise and serrated twisted tape with twist angle 30°, 60° and 90° were experimented in the tube for Reynolds number 4000–20,000. They found that modified clockwise and counterclockwise twisted tape generated periodic change in swirl direction causing better mixing as well as it increased the heat transfer rate up to 72.2%.

Tabatabaeikia et al. (2014) reviewed the effective techniques for achieving high heat transfer in heat exchanger. The enhanced heat transfer rate provides compactness, energy saving, cost reduction, light weight and small heat exchangers. From reviewing different research papers, they concluded that serrated twisted tapes and peripherally cut twisted tapes increased the mean heat transfer rate up to 72%. They observed that V-cut twisted tape raised the rate of heat transfer to 10% in comparison to plain twisted tube at the similar conditions. They reviewed works on various kinds of typical twisted tapes, perforated twisted tapes, notched twisted tapes, jagged twisted tapes and butterfly inserts.

The Nusselt number and thermo-hydraulic performance were compared, and it was found that jagged insert performance was best followed by that of classic twisted tapes, then perforated twisted tape and notched twisted tapes. They observed 40% higher Nusselt number associated with jagged TT in comparison to perforated TT. Similarly, for helical screw tape, 12% increase in heat transfer rate was found as twist ratio increased. Twisted tape with alternate axis was found better than that of classic twisted tapes. The propeller-type vortex generator achieved 18–163% increase in heat transfer as compared to plain tube. Finally, they concluded that combination of various types of inserts with roughness increased the heat transfer rate significantly but the pressure loss also increased.

References

Abolarin SM, Everts M, Meyer JP (2019) Heat transfer and pressure drop characteristics of alternating clockwise and counter clockwise twisted tape inserts in the transitional flow regime. Int J Heat Mass Transf 133:203–217

Bandyopadhyay PS, Gaitonde UN, Sukhatme SP (1991) Influence of free convection on heat transfer during laminar flow in tubes with twisted tapes. Exp Thermal Fluid Sci 4(5):577–586

Bas H, Ozceyhan V (2012) Heat transfer enhancement in a tube with twisted tape inserts placed separately from the tube wall. Exp Thermal Fluid Sci 41:51–58

Bergles AE, Joshi SD (1983) Augmentation techniques for low Reynolds number in-tube flow. In: Low Reynolds number flow heat exchangers. Hemisphere, Washington, DC, pp 694–720

Bhattacharyya S, Saha SK (2012) Thermohydraulic of laminar flow through a circular tube having integral helical rib roughness and fitted with centre-cleared twisted-tape. Exp Thermal Fluid Sci 42:154–162

Bhuiya MMK, Chowdhury MSU, Saha M, Islam MT (2013) Heat transfer and friction factor characteristics in turbulent flow through a tube fitted with perforated twisted tape inserts. Int Commun Heat Mass Transf 46:49–57

Boonloi A, Jedsadaratanachai W (2016) Turbulent forced convection and heat transfer characteristic in a circular tube with modified-twisted tapes. Int J Thermodyn 2016. Article ID 8235375, 16 pages

Chamoli S, Yu P, Yu SM (2017) Multi-objective shape optimization of a heat exchanger tube fitted with compound inserts. Appl Therm Eng 117:708–724

Chakroun WM, Al-Fahed SF (1996) The effect of twisted-tape width on heat transfer and pressure drop for fully developed laminar flow. J Eng Gas Turb Power 118(3):584–589

Chang SW, Guo MH (2012) Thermal performances of enhanced smooth and spiky twisted tapes for laminar and turbulent tubular flows. Int J Heat Mass Transf 55(25–26):7651–7667

Chen L, Zhang HJ (1993) Convection heat transfer enhancement of oil in a circular tube with spiral spring inserts. In: Chow LC, Emery AF (eds) Heat transfer measurements and analysis, HTD-ASME Symp., vol 249, pp 45–50

Coetzee H, Liebenberg L, Meyer JP (2001) Heat transfer and pressure drop characteristics of angled spiralling tape inserts in a heat exchanger annulus. In: Paper RA, Aminemi NK, Toma O, Rudland R, Crain E (eds) Proceedings of the ASME process industries division. ASME, New York, pp 79–86

Chokphoemphun S, Pimsarn M, Thianpong C, Promvonge P (2015) Thermal performance of tubular heat exchanger with multiple twisted-tape inserts. Chin J Chem Eng 23:755–762

Date AW (1974) Prediction of fully-developed flow containing a twisted-tape. Int J Heat Mass Transfer 17:845–859

Date AW (1973) Flow in tubes containing twisted tapes. Heat Vent Eng 47:240–249

Date AW, Saha SK (1990) Numerical prediction of laminar flow and heat transfer characteristics in a tube fitted with regularly spaced twisted-tape elements. Int J Heat Fluid Flow 11(4):346–354

Date AW, Singham JR (1972) Numerical prediction of friction and heat transfer characteristics of fully developed laminar flow in tubes containing twisted tapes, ASME Paper 72-HT-l 7. ASME, New York

Duplessis JP, Kroeger DG (1983) Numerical prediction of laminar flow with heat transfer in tube with a twisted tape insert. In: Proceedings of the international conference on numerical methods in laminar and turbulent flow, pp 775–785

Eiamsa-ard S (2010) Study on thermal and fluid flow characteristics in turbulent channel flows with multiple twisted tape vortex generators. Int Commun Heat Mass Transf 37:644–651

Eiamsa-ard S, Promvonge P (2010a) Performance assessment in a heat exchanger tube with alternate clockwise and counter-clockwise twisted-tape inserts. Int J Heat Mass Transf 53 (7–8):1364–1372

Eiamsa-ard S, Promvonge P (2010b) Thermal characteristics in round tube fitted with serrated twisted tape. Appl Therm Eng 30:1673–1682

Eiamsa-ard S, Thianpong C, Eiamsa-ard P, Promvonge P (2010a) Thermal characteristics in a heat exchanger tube fitted with dual twisted tape elements in tandem. Int Commun Heat Mass Transf 37:39–46

Eiamsa-ard S, Wongcharee K, Eiamsa-ard P, Thianpong C (2010b) Thermohydraulic investigation of turbulent flow through a round tube equipped with twisted tapes consisting of centre wings and alternate-axes. Exp Thermal Fluid Sci 34:1151–1161

Eiamsa-ard S, Thianpong C, Eiamsa-ard Turbulent P (2010c) Heat transfer enhancement by counter co-swirling flow in a tube fitted with twin twisted tapes. Exp Thermal Fluid Sci 34:53–62

Eiamsa-ard S, Wongcharee K, Eiamsa-ard P, Thianpong C (2010d) Heat transfer enhancement in a tube using delta-winglet twisted tape inserts. Appl Therm Eng 30:310–318

Eiamsa-ard S, Seemawute P, Wongcharee K (2010e) Influences of peripherally-cut twisted tape insert on heat transfer and thermal performance characteristics in laminar and turbulent tube flows. Exp Thermal Fluid Sci 34:711–719

Eiamsa-ard S, Nanan K, Thianpong C, Eiamsa-ard P (2013) Thermal performance evaluation of heat exchanger tubes equipped with coupling twisted tapes. Exp Heat Transf 26:413–430

Eiamsa-ard S, Kiatkittipong K, Jedsadaratanachai W (2015a) Heat transfer enhancement of TiO_2/water nanofluid in a heat exchanger tube equipped with overlapped dual twisted-tapes. Int J Eng Sci Tech 18:336–350

Eiamsa-Ard S, Nanan K, Wongcharee K, Yongsiri K, Thianpong C (2015b) Thermohydraulic performance of heat exchanger tube equipped with single-, double-, and triple-helical twisted tapes. Chem Eng Commun 202:606–615

Goudarzi K, Jamali H (2017) Heat transfer enhancement of Al_2O_3-EG nanofluid in a car radiator with wire coil inserts. Appl Therm Eng 118:510–517

Gupte N, Date AW (1989) Friction and heat transfer characteristics of helical turbulent air flow in annuli. J Heat Transf 111:337–344

Hindasageri V, Vedula RP, Prabhu SV (2015) Heat transfer distribution of swirling flame jet impinging on a flat plate using twisted tapes. Int J Heat Mass Transf 91:1128–1139

Hong SW, Bergles AE (1976) Augmentation of laminar flow heat transfer in tubes by means of twisted tape inserts. J Heat Transf 98:251–256

Holman JP (2009) Heat transfer [M], 10th edn. McGraw-Hill Education, New York

Hong YX, Deng XH, Zhang LS (2012) 3D numerical study on compound heat transfer enhancement of converging-diverging tubes equipped with twin twisted tapes. Chin J Chem Eng 20:589–601

Hong Y, Du J, Wang S (2017) Turbulent thermal, fluid flow and thermodynamic characteristics in a plain tube fitted with overlapped multiple twisted tapes. Int J Heat Mass Transf 115:551–565

Jaisankar S, Radhakrishnan TK, Sheeba KN, Suresh S (2009a) Experimental investigation of heat transfer and friction factor characteristics of thermosyphon solar water heater system fitted with spacer at the trailing edge of left–right twisted tapes. Energy Convers Manag 50:2638–2649

Jaisankar S, Radhakrishnan TR, Sheeba KN (2009b) Studies on heat transfer and friction factor characteristics of thermosyphon solar water heating system with helical twisted tapes. Energy 34:1054–1064

Kanizawa FT, Mogaji TS, Ribatski G (2014) Evaluation of the heat transfer enhancement and pressure drop penalty during flow boiling inside tubes containing twisted tape insert. Appl Therm Eng 70(1):328–340

Klepper O (1972) Heat transfer performance of short twisted tapes. Oak Ridge National Lab, Oak Ridge

Kumar A, Prasad BN (2000) Investigation of twisted tape inserted solar water heaters—heat transfer, friction factor and thermal performance results. Renew Energy 19(3):379–398

Kumar NR, Bhramara P, Sundar LS, Singh MK, Sousa AC (2017) Heat transfer, friction factor and effectiveness of Fe_3O_4 nanofluid flow in an inner tube of double pipe U-bend heat exchanger with and without longitudinal strip inserts. Exp Thermal Fluid Sci 85:331–343

Lekurwale RA, Ingole PR, Joshi YG, Ingole PR (2014) Performance assessment of heat exchanger tubes to improve the heat transfer rate in turbulant flows by using different types of twisted tapes inserts in tubes

Lepina RF, Bergles AE (1969) Heat transfer and pressure drop in tape-generated swirl flow of single-phase water. ASME J Heat Transf 91:434–442

Liu X, Li C, Cao X, Yan C, Ding, M (2018) Numerical analysis on enhanced performance of new coaxial cross twisted tapes for laminar convective heat transfer. Int J Heat Mass Transf 121:1125–1136

Maddah H, Alizadeh M, Ghasemi N, Wan Alwi SR (2014) Experimental study of Al_2O_3/water nanofluid turbulent heat transfer enhancement in the horizontal double pipes fitted with modified twisted tapes. Int J Heat Mass Transf 78:1042–1054

Manglik RM (1991) Heat transfer enhancement of in-tube flows in process heat exchangers by means of twisted-tape inserts. Ph.D. thesis, Department of Mechanical Engineering, Rensselaer Polytechnic Institute, Troy, NY

Manglik RM, Bergles AE (1992a) Heat transfer and pressure drop correlations for twisted-tape inserts in isothermal tubes: part I, laminar flows. In Pate MB, Jensen MK (eds) Enhanced heat transfer, ASME Symp. HTD, vol 202, pp 89–98

Manglik RM, Bergles AE (1992b) Heat transfer and pressure drop correlations for twisted-tape inserts in isothermal tubes: part II, transition and turbulent flows. In: Pate MB, Jensen MK (eds) Enhanced heat transfer ASME Symp. HTD, vol 202, pp 99–106

Manglik RM, Bergles AE (1993a) Heat transfer and pressure drop correlations for twisted-tape inserts in isothermal tubes. Part I- Laminar flows J. Heat Transfer 115:881–889

Manglik RM, Bergles AE (1993b) Heat transfer and pressure drop correlations for twisted-tape inserts in isothermal tubes: Part II-Transition and turbulent flows J. Heat Transfer 115:890–896

Marner WJ, Bergles AE (1978) Augmentation of tube-side laminar flow heat transfer by means of twisted tape inserts, static mixer inserts and internally finned tubes. In: Heat transfer 1978, Proc. 6th international heat transfer conference, vol 2. Hemisphere, Washington, DC, pp 583–588

Marner WJ, Bergles AE (1985) Augmentation of highly viscous laminar tube side heat transfer by means of a twisted-tape insert and an internally finned tube. In: Advances in enhanced heat transfer—1985

Meyer JP, Abolarin SM (2018) Heat transfer and pressure drop in the transitional flow regime for a smooth circular tube with twisted tape inserts and a square edged inlet. Int J Heat Mass Transf 117:11–29

Noothong W, Eiamsa-ard S, Promvonge P (2006) Effect of twisted-tape inserts on heat transfer in a tube. In: The 2nd joint international conference on sustainable energy and environment SEE, pp 1–5

Patil AG (2000) Laminar flow heat transfer and pressure drop characteristics of power-law fluids inside tubes with varying width twisted tape inserts. J Heat Transf 122:143–149

Piriyarungrod N, Eiamsa-Ard S, Thianpong C, Pimsarn M, Nanan K (2015) Heat transfer enhancement by tapered twisted tape inserts. Chem Eng Process Process Intensif 96:62–71

Promvonge P (2015) Thermal performance in square-duct heat exchanger with quadruple V-finned twisted tapes. Appl Therm Eng 91:298–307

Qi C, Liu M, Luo T, Pan Y, Rao Z (2018) Effects of twisted tape structures on thermo-hydraulic performances of nanofluids in a triangular tube. Int J Heat Mass Transf 127:146–159

Ray S, Date AW (2003) Friction and heat transfer characteristics of flow through square duct with twisted tape insert. Int J Heat Mass Transf 46(5):889–902

Rios-Iribe EY, Cervantes-Gaxiola ME, Rubio-Castro E et al (2015) Heat transfer analysis of a non-Newtonian fluid flowing through a circular tube with twisted tape inserts. Appl Therm Eng 84:225–236

Safikhani H, Abbasi F (2015) Numerical study of nanofluid flow in flat tubes fitted with multiple twisted tapes. Adv Powder Technol 26(6):1609–1617

Saha SK, Gaitonde UN, Date AW (1989) Heat transfer and pressure drop characteristics of laminar flow in a circular tube fitted with regularly spaced twisted-tape elements. Exp Therm Fluid Sci 2:310–322

Saha SK, Dutta A, Dhal SK (2001) Friction and heat transfer characteristics of laminar swirl flow through a circular tube fitted with regularly spaced twisted-tape elements. Int J Heat Mass Transf 44:4211–4223

Salam B, Biswas S, Saha S, Bhuiya MMK (2013) Heat transfer enhancement in a tube using rectangular-cut twisted tape insert. Procedia Eng 56:96–103

Salman SD, Kadhum AAH, Takriff MS, Mohamad AB (2014) Heat transfer enhancement of laminar nanofluids flow in a circular tube fitted with parabolic-cut twisted tape inserts. Sci World J 2014. Article ID 543231, 7 pages

Sarada N, Raju S, Kalyani Radha AV, Shyam K (2010) Enhancement of heat transfer using varying width twisted tape inserts. Int J Eng Sci Technol 2(6):107–118

Seemawute P, Eiamsa-ard S (2010) Thermohydraulics of turbulent flow through a round tube by a peripherally-cut twisted tape with an alternate axis. Int Commun Heat Mass Transf 37:652–659

Shivkumar C, Rao MR (1998) Studies on compound augmentation of laminar flow heat transfer to generalized power law fluids in spirally corrugated tubes by means of twisted tape inserts. In: Jacobs HR (ed) ASME Proc. 96, 1988 national heat transfer conference, Vol. 1, HTD, 96, pp 685–692

Smithberg E, Landis F (1964) Friction and forced convection heat transfer characteristics in tubes with twisted tape swirl generators. J Heat Transf 87:39–49

Sivakumar K, Rajan K, Murali S, Prakash S, Thanigaivel V, Suryakumar T (2015) Experimental investigation of twisted tape insert on laminar flow with uniform heat flux for enhancement of heat transfer. J Chem Pharm Sci 201:205

Sukhatme SP, Gaitonde UN, Shidore CS, Kuncolienkar RS (1987) Forced convection heat transfer to a viscous liquid in laminar flow in a tube with a twisted-tape. In: Paper No. HMT-87, part B, proceeding of the 9th national heat and mass transfer conference, IISc, Bangalore, pp 1–3

Sundar LS, Bhramara P, Kumar NR, Singh MK, Sousa AC (2017) Experimental heat transfer, friction factor and effectiveness analysis of Fe3O4 nanofluid flow in a horizontal plain tube with return bend and wire coil inserts. Int J Heat Mass Transf 109:440–453

Sundar LS, Singh MK, Sousa AC (2018) Heat transfer and friction factor of nanodiamond-nickel hybrid nanofluids flow in a tube with longitudinal strip inserts. Int J Heat Mass Transf 121:390–401

Tabatabaeikia S, Mohammed HA, Nik-Ghazali N, Shahizare B (2014) Heat transfer enhancement by using different types of inserts. Adv Mech Eng 6:250354

Tamna S, Kaewkohkiat Y, Skullong S, Promvonge P (2016) Heat transfer enhancement in tubular heat exchanger with double V-ribbed twisted-tapes case studies. Therm Eng 7:14–24

Thianpong C, Eiamsa-Ard P, Promvonge P, Eiamsa-Ard S (2012) Effect of perforated twisted-tapes with parallel wings on heat transfer enhancement in a heat exchanger tube. Energy Procedia 14:1117–1123

Thorsen R, Landis F (1968) Friction and beat transfer characteristics in turbulent swirl flow subjected to large transverse temperature gradients. J Heat Transf 90:87–89

Webb RL, Kim NH (2005) Principles of enhanced heat transfer. Taylor & Francis, New York

Wongcharee K, Eiamsa-ard S (2011) Friction and heat transfer characteristics of laminar swirl flow through the round tubes inserted with alternate clockwise and counter-clockwise twisted-tapes. Int Commun Heat Mass Transf 38:348–352

Xie L, Gu R, Zhang X (1992) A study of the optimum inserts for enhancing convective heat transfer of high viscosity fluid in a tube. In: Chen X-J, Vezirogln TN, Tien CL (eds) Multiphase flow and heat transfer; Second international symposium, vol 1. Hemisphere, New York, pp 649–656

Yakovlev AB, Tarasevich SE (2014) Heat transfer and hydraulic resistance of forced convection in tubes with twisted-tape inserts and different inlet conditions. J Enhanc Heat Transf 21 (2–3):183–194

Zhang Z, Ma D, Fang X, Gao X (2008) Experimental and numerical heat transfer in a helically baffled heat exchanger combined with one three-dimensional finned tube. Chem Eng Process 47:1738–1743

Zhang YF, Li FY, Liang ZM (1991) Heat transfer in spiral-coil-inserted tubes and its application. In: Ebadian MA, Pepper DW, Diller T (eds) Advances in heat transfer augmentation, ASME Symp. HTD, vol 169, pp 31–36

Zhang YM, Han JC, Lee CP (1997) Heat transfer and friction characteristics of turbulent flow in circular tubes with twisted-tape inserts and axial interrupted ribs. J Enhanc Heat Transf 4(4):297

Zhuo N, Ma QL, Zhang ZY, Sun JQ, He J (1992) Friction and heat transfer characteristics in a tube with a loose fitting twisted-tape insert. In: Chen X-J, Veziroglu TN, Tien CL (eds) Multiphase flow and heat transfer. Second international symposium, vol 1. Hemisphere, New York, pp 657–661

Zimparov V (2001) Enhancement of heat transfer by a combination of three-start spirally corrugated tubes with a twisted tape. Int J Heat Mass Transf 44:551–574

Zimparov V (2002) Enhancement of heat transfer by a combination of a single-start spirally corrugated tubes with a twisted tape. Exp Therm Fluid Sci 25:535–546

Zimparov V (2004a) Prediction of friction factors and heat transfer coefficients for turbulent flow in corrugated tubes combined with twisted tape inserts. Part I: friction factors. Int J Heat Mass Transf 47:589–599

Zimparov V (2004b) Prediction of friction factors and heat transfer coefficients for turbulent flow in corrugated tubes combined with twisted tape inserts. Part 2: heat transfer coefficients. Int J Heat Mass Transf 47:385–393

Chapter 3
Displaced Enhancement Devices and Wire Coil Inserts

3.1 Displaced Enhancement Devices

These devices periodically mix the gross flow structure and accelerate the local velocity near the wall (Koch 1958; Evans and Churchill 1963; Colburn and King 1931; Megerlin et al. 1974; Bergles 1985). These devices provide very poor performance, and the velocity in the enhanced tube must be reduced to a very low value.

Xie et al. (1992) worked with interrupted elements, and they gave the correlations

$$Nu_d = 0.0202 Re_{d,f}^{0.767} Pr^{2/3} y^{-0.282} \left(\frac{\mu}{\mu_w}\right)^{0.14} \tag{3.1}$$

$$f = \left(5.68 - 2.07y + 0.43y^2\right) Re_{d,f}^{-\left(0.34-1.41y-0.021y^2\right)} (\mu/\mu_w) 1/3 \tag{3.2}$$

Park et al. (2000) worked with air for $20{,}000 < Re_d < 80{,}000$ (Fig. 3.1). Additional data on straight or cross tapes for turbulent airflow are reported by Hsieh and Kuo (1994).

Displaced enhancement devices are better suited to laminar than turbulent flow, since the thermal impedance is not confined to a thin boundary layer region near the wall in laminar flow (Oliver and Aldington 1986).

Ahire et al. (2017) carried out an experiment on heat exchanger by passive technique using conical ring insert. Test fluid flows in circular pipe had been subjected to constant heat flux. They used copper and aluminium as conical ring insert material. They aimed to find out heat transfer coefficient, friction factor, pressure drop and heat transfer rate by using various conical rings with varying pitches and different materials. The experimental results with conical insert were compared with the results of smooth tubes. Figures 3.2 and 3.3 show the variation of Nusselt number with Reynolds number for aluminium and copper conical ring inserts, respectively. Figures 3.4 and 3.5 show the comparison of friction factors

© The Author(s), under exclusive license to Springer Nature Switzerland AG 2020
S. K. Saha et al., *Insert Devices and Integral Roughness in Heat Transfer Enhancement*, SpringerBriefs in Applied Sciences and Technology,
https://doi.org/10.1007/978-3-030-20776-2_3

Fig. 3.1 h/h_P vs. Re (FG-2a PEC analysis) for various types of twisted tapes for air in an electrically heated tube with $20{,}000 \leq Re \leq 80{,}000$ (from Park et al. 2000)

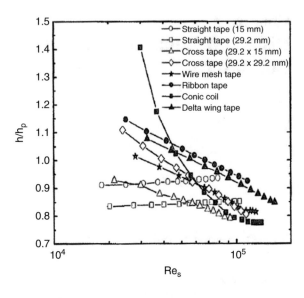

Fig. 3.2 Comparison of Nusselt number vs. Reynolds number for aluminium inserts (Durgesh et al. 2017)

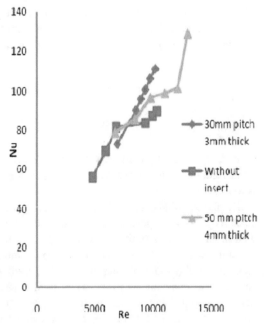

with Reynolds number for two pitches and thickness of aluminium insert and copper insert, respectively. They observed that heat transfer coefficient increases and friction factors decreases with increase in Reynolds numbers. The results showed that smaller pitch and thickness of conical ring insert gave optimum result for greater Nusselt number.

Fig. 3.3 Comparison of
Nusselt
number vs. Reynolds
number for copper inserts
(Durgesh et al. 2017)

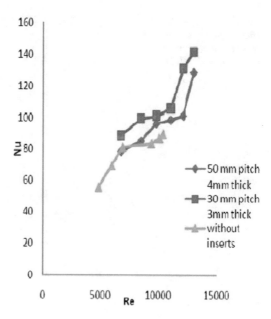

Fig. 3.4 Comparison of
friction factor vs. Reynolds
number for aluminium
inserts (Durgesh et al. 2017)

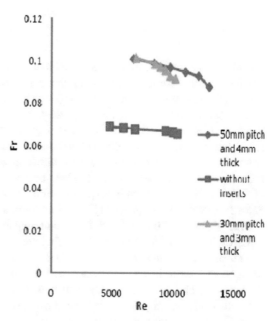

Tu et al. (2015) numerically investigated the convective heat transfer performance in circular tube by using mesh cylinder inserts. They studied the effect of open area rate, inlet closed type, spacer length and insert length on the heat transfer enhancement in the Reynolds number range of 280–1800. They observed that

Fig. 3.5 Comparison of
friction factor vs. Reynolds
number for copper inserts
(Durgesh et al. 2017)

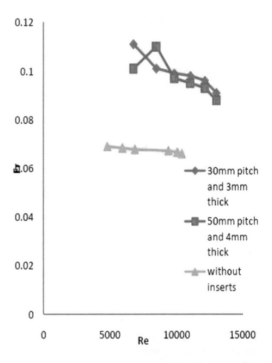

boundary layer redevelopment leads to increase in Nusselt number at the inlet and
annular region of complex laminar regime. A small hole was cut in the closed inlet of
mesh cylinder inserts which helped to reduce the flow resistance without decrement
in the heat transfer rate. Figure 3.6 presents the variation of Nusselt numbers as a
function of Reynolds number for insert with two types of inlet closed. Numerical
simulated results reported that friction factor was increased up to 10–13 times
compared to the smooth tube due to which flow resistance becomes high. The
high flow resistance deteriorated the heat transfer rate. Figure 3.7 shows variation
of flow friction factor with Reynolds number for mesh insert with the sample
Sec-Close (with closed inlet) and Sec-Hole (with a small hole at the inlet).

Abu-Mulaweh (2003) carried out an experimental study to compare four different
types of heat transfer enhancement techniques: two insert devices (displacement
device and swirl flow device), extended surfaces and obstruction devices. He
reported that these insert devices assisted to reduce the heat transfer surface area
for the same amount of heat duty, and thus, both size of heat exchanger and cost
reduced. These techniques also reduced the logarithmic mean temperature difference
(LMTD) and pumping power for fixed heat duty and surface area. Automobile,
refrigeration industries, process industries, etc. routinely use enhanced heat transfer
surfaces in their heat exchanger. The displacement device was basically a spiralled
rod. This device was inserted into the flow channel to indirectly improve heat
transfer by promoting turbulence and flow mixing.

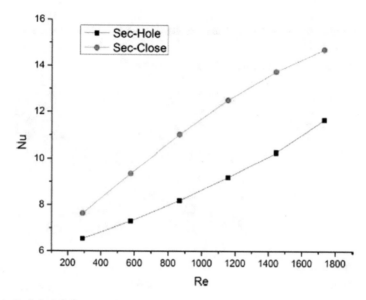

Fig. 3.6 Variation of *Nu* vs. *Re* for mesh inserts with different inlet closed (Tu et al. 2015)

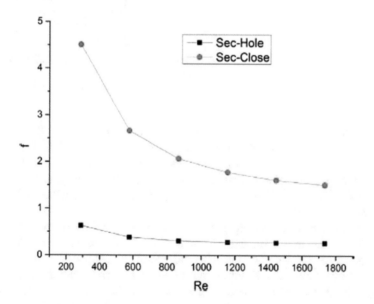

Fig. 3.7 Variation of *f* vs. *Re* for mesh inserts with different inlet closed (Tu et al. 2015)

Secondary flow developed flow field inside the annulus. Spiralled rod reduced the hydraulic diameter of the heat exchanger. Figure 3.8 shows the stainless steel S316 spiralled rod having 0.125 in. diameter with pins inserted. He used an extended surface to enhance the heat transfer rate in the heat exchangers. A total of 42 copper

Fig. 3.8 Illustration of the spiralled rod (Abu-Mulaweh 2003)

Fig. 3.9 Copper tube with
the attached fins and a
portion of shell removed
(Abu-Mulaweh 2003)

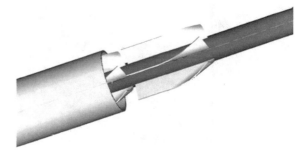

Fig. 3.10 Twisted tape
inside the copper tube
(Abu-Mulaweh 2003)

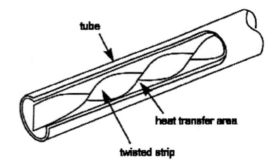

fins were attached on the inner tube and placed in seven banks of six fins each. The
length of each bank of fins was kept 2 in. Fin height of 7/16 and fin thickness of
0.025 in. were chosen. The fin was made of copper due to its high thermal
conductivity. Figure 3.9 shows the inner copper tube with attached fins and portion
of the shell removed. He also carried out an experiment with thin twisted helical tape,
as shown in Fig. 3.10.

The twisted tape having eight twists was made of an aluminium sheet with
dimension of 1.6 mm thick and 12.7 mm wide. Small clearance should be kept
between the twisted tape and the inner surface of the copper tube to allow easy
insertion of the tape. Manglik and Bergles (1992a, b) worked with twisted tape for
finding the heat transfer and pressure drop correlation in both laminar and turbulent
flow. Figure 3.11 shows the performance of the four modified heat exchangers based
on heat duty and mass flow rate. The conditions of experimental results were:
counter-flow, mass flow rate of cooling water of 10 g/s, cooling water and hot
water inlet temperature of 20 °C and 70 °C, respectively. Among the four enhancement techniques, annular disk and the spiral rod showed highest and lowest heat
transfer rate, respectively.

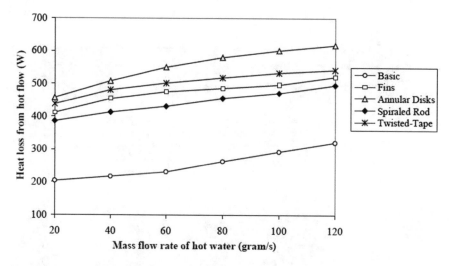

Fig. 3.11 Performance of the modified heat exchangers (Abu-Mulaweh 2003)

Liu et al. (2018) carried out an experimental and numerical analysis on the laminar flow heat transfer characteristics of a heat exchanger circular tube fitted with multiple conical strip inserts. They investigated the effect of number of conical strips (n), central angle (α), slant angle (θ) and pitch (p). They observed that heat transfer rate and friction factor increased with increasing number of conical strips and central angle and decreasing pitch. Nusselt number and flow resistance increased first and then decreased with increase of slant angle. Both heat transfer and friction factor were increased by approximately 2.54–7.63 and 2.40–28.74 times, respectively, compared to the plain tube. The result of numerical simulation revealed that the overall heat transfer performance lied in the range of 1.23–6.05. Manglik and Bergles (1993a, b), Saha et al. (1989), Guo et al. (2011), Saysroy and Eiamsa-ard (2017), Zhang et al. (2013), Fan et al. (2012), Tu et al. (2015) and Sivashanmugam and Suresh (2006) also investigated the effect of various types of inserts on the performance of heat transfer rate.

Heat transfer enhancement in flow boiling in an annular tube has been studied by Wen et al. (2015) using metallic porous inserts. They have addressed the effect of perforated copper porous inserts on boiling of R600a refrigerant. The effect of mass velocity, vapour quality and geometrical parameters of the tube on heat transfer has been studied. The test channels with dispersed copper porous inserts have been shown in Fig. 3.12. The geometrical details of the porous channels have been shown in Table 3.1. Figure 3.13 shows the variation of heat transfer coefficient enhancement ratio with vapour quality for a given mass velocity and two different heat fluxes.

The variation of heat transfer coefficient corresponding to the variation of vapour quality for a fixed heat flux and two different mass velocities has been shown in Fig. 3.14. The pressure drop ratio for tube with inserts and without inserts has been

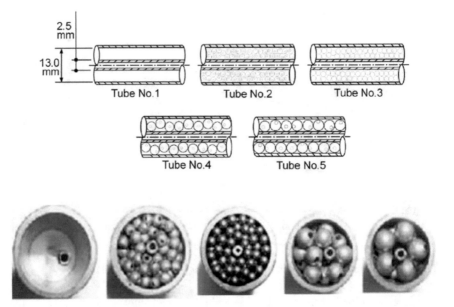

Fig. 3.12 Test channels with dispersed copper porous inserts (Wen et al. 2015)

Table 3.1 Geometrical details of the porous channels (Wen et al. 2015)

Tube no.	Diameter of copper, D_p (mm)	Diameter of hollow (mm)	Surface area density (mm^2/mm^3)	Mean pore diameter, D_e (mm)	Porosity	Permeability (m^2)
1	**Open tube**					
2	2	0.9	6.25	0.293	0.181	1.31E − 09
3	2	0	3.0	0.137	0.093	0.02E − 09
4	4	1.1	2.16	0.386	0.224	1.67E − 09
5	5	1.5	1.81	0.475	0.263	4.63E − 09

plotted in Fig. 3.15 against vapour quality. They concluded that the effect of vapour quality on heat transfer coefficient enhancement ratio was insignificant. As the surface area density of tube 2 is higher due to hollow, the enhancement ratio for tube 2 was greater than that for tube 3 which does not have a hollow. The increase in heat transfer coefficient for tubes having inserts with respect to smooth tube are 1.23–1.36, 1.10–1.20, 1.13–1.22 and 1.18–1.28 for tubes 2, 3, 4 and 5, respectively.

Ozceyhan et al. (2008) numerically simulated two-dimensional domain and modelled a long tube in which fully developed flow can develop. They evaluated the heat transfer enhancement for the tube consisting of circular cross-sectioned ring. They considered clearance between wall and the ring and computed for five distinct locations for rings that are located at $P = d/2$, d, $3d/2$, $2d$ and $3d$. The assumptions used were fully developed steady turbulent flow with fixed thermal conductivity of tube wall and was homogenous and isotropic. The objective of the study was to

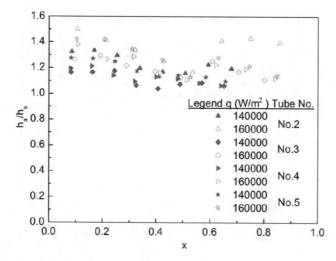

Fig. 3.13 Variation of heat transfer coefficient enhancement ratio with vapour quality for a given mass velocity and two different heat fluxes (Wen et al. 2015)

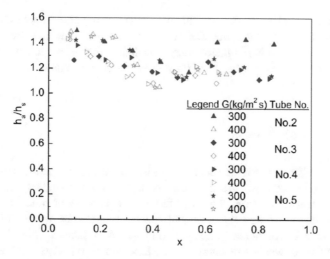

Fig. 3.14 Variation of heat transfer coefficient corresponding to the variation of vapour quality (Wen et al. 2015)

predict the influence of ring spacing and distance from wall on heat transfer characteristics.

They concluded from the simulated results and validated it with previous work and found that Nusselt number raised and friction factor decreased with increased Reynolds number. It was highest at $P = d/2$. They found that both Nusselt number and friction factor increased as ring spacing decreased. Results revealed that $P = d/2$-type inserts induces maximum pressure drop than those available types. It was

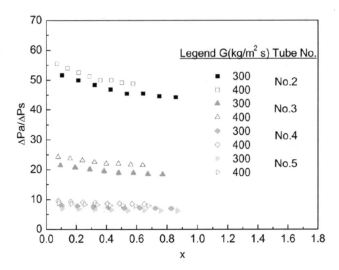

Fig. 3.15 Pressure drop ratio for tube with inserts and without inserts (Wen et al. 2015)

observed that overall enhancement ratio was greater than unity except for $d/2$ where friction overcomes the heat transfer. Also, enhancement ratio increased with the increase in ring spacing. Finally, they concluded that 18% enhancement was obtained for Reynolds number 15,000 for in-between ring spacing.

3.2 Wire Coil Inserts

These types of inserts provide enhancement by flow separation at the wire, causing fluid mixing in the downstream boundary layer. Since boundary layer mixing dissipates downstream from the wire, the local enhancement quickly dissipates. Rational selection of wire diameter and spacing requires knowledge of how the local heat transfer coefficient varies with dimensionless distance (p/e) downstream from the wire (Edwards and Sheriff 1961; Emerson 1961) (Figs. 3.16 and 3.17). Layer wire diameter protrudes into the turbulence-dominated boundary layer, and no additional benefit is obtained. On the contrary, larger wires have increased profile drag, and this causes increased momentum loss and pressure drop. The wire coil insert is wall-attached roughness of special helical ribs (Uttarwar and Raja Rao 1985; Chen and Zhang 1993; Sethumadhavan and Raja Rao 1983).

$$Nu_{Dv} = 1.65 \tan \alpha Re_{Dv}{}^{m} Pr^{0.35} \left(\frac{\mu}{\mu_{w}}\right)^{0.14} \tag{3.3}$$

Fig. 3.16 Enhancement of the local heat transfer coefficient provided by a 1.58-mm-diameter transverse wire for boundary layer flow of air over a flat plate at $ReL = 530{,}000$ (from Edwards and Sheriff 1961)

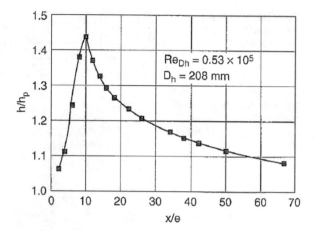

Fig. 3.17 Effect of Re_1 and wire diameter on h_{max}/h_P for boundary layer flow of air over a flat plate for wire sizes between 0.40 and 6.35 mm (from Edwards and Sheriff 1961)

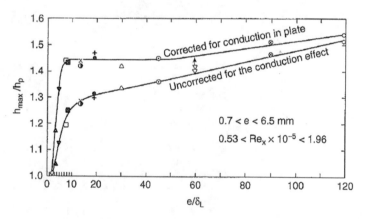

$$Nu_d = 1.258Re_d^{0.566}Pr^{0.169}(p/d_i)^{0.186}(p/e)^{-0.408} \tag{3.4}$$

$$f = 95.049Re_d^{-0.129}Pr^{-0.230}(p/d_i)^{0.848}(p/e)^{-1.428} \tag{3.5}$$

Herring and Heister (2009) carried out an experiment to improve the performance of compact heat exchanger by using wire coil inserts in a tube. They used JP-10 fuel in electrically heated tube at relevant temperature and performed CFD calculations. A Nusselt number correlation was developed, and the results showed that wire coils having three different pitches enhanced the heat transfer rate. Table 3.2 shows the dimensional parameters of wire coils that were used in the experimental test. Figure 3.18 shows that the local Nusselt numbers varied with Reynolds numbers and compares all wire coils with plain tube. Bruening and Chang (1999), Garcia et al. (2005) and Inaba and Ozaki (1997) worked on cooling system by enhancing heat transfer rate using wire coil insert.

Table 3.2 Wire coil parameters (Herring and Heister 2009)

D_c (in)	e (in)	e/D_i	α	P/D_i	P/e
0.094	0.014	0.203	47.4°	2.9	14.3
0.120	0.014	0.203	33.5°	4.8	23.6
0.156	0.014	0.203	26.3°	6.4	31.5

Fig. 3.18 Variation of local Nusselt number for all the tests (Herring and Heister 2009)

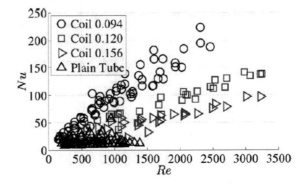

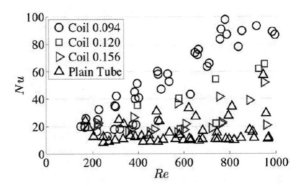

Khatua et al. (2014) studied the condensation phenomenon of refrigerant R245fa in a tube fitted with coiled wire inserts. They focussed on the pressure drop in the tube with coiled wire inserts. They observed significant effect of vapour quality, mass flow velocity and the pitch of the coil on pressure drop. The geometry of the coiled wires has been shown in Table 3.3. The effect of vapour quality on pressure drop has been shown in Fig. 3.19 for coiled wires having different pitch. They observed greater pressure drop in tubes with coiled wire compared to that in plain tube. They reported that the higher pressure drop in tube with coiled wire was due to the centrifugal force created by the flow over coiled wire. The pressure drop in tube with different coiled wires has been observed to be 14.59–13.7 times greater than that in plain tube. The variation of ratio of pressure drop in tube with coiled wires and plain tube with Reynolds number has been presented in Fig. 3.20.

The pressure drop ratio was observed to be increased with decrease in coil pitch. This is because, as the coil pitch decreases, the turbulence increases leading to increased induced swirl flow. They have also observed that at higher heat fluxes,

Table 3.3 Geometry of the coiled wires (Khatua et al. 2014)

Serial number	Wire diameter, e (mm)	Coil pitch, P_c (mm)	Helix angle, α (deg)	Spring coil diameter, D (mm)
1	1.01	3.0	72	3.98
2	1.01	6.3	55.9	3.97
3	1.01	8.4	47.9	3.96
4	1.01	11.0	40.2	3.95
5	1.01	14.0	33.6	3.94
6	1.01	16.6	29.3	3.93

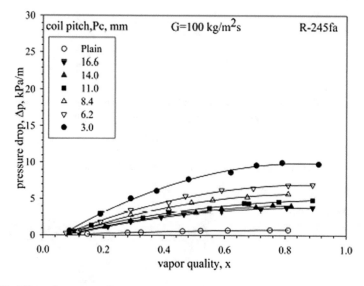

Fig. 3.19 Effect of vapour quality on pressure drop (Khatua et al. 2014)

the effect of coil pitch on pressure drop was negligible. They concluded that the use of coiled wires in high vapour quality region results in greater pressure drop than that in low vapour quality region. The highest value of pressure drop was observed to be 13.33 for coiled wire with 3 mm pitch at 150 kg/m^2 s mass flux. On the other hand, the tube with coiled wire having 16.6 mm pitch showed the minimum pressure drop (4.12) at 250 kg/m^2 s mass flux. Similar works on pressure drop in condensation process using different inserts have been presented by Hejaji et al. (2010), Dewan et al. (2004), Naphon (2006), Said and Azer (1983), Akhavan-Behabadi et al. (2008), Ould Didi et al. (2002) and Agrawal et al. (2004).

The performance of twisted wire inserts in annular channels was studied by Yakovlev (2013). The wire was taken such that the diameter of the wire cross section is equal to the height of the annular gap. The annular tube with twisted wire has been shown in Fig. 3.21. The annular channels with continuous twisting on length have been considered for their study. The continuous flow twisting can be generally

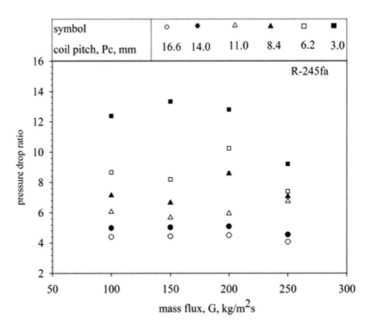

Fig. 3.20 Variation of ratio of pressure drop in tube with coiled wires and plain tube (Khatua et al. 2014)

Fig. 3.21 Annular tube with twisted wire (Yakovlev 2013)

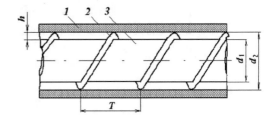

achieved by using a wire which is spirally wound on the central body. They observed that the heat transfer rate on the concave surface of the annular channel with continuous twisting was comparatively more than that on the concave surface. They concluded that the pressure drop calculations in annular channels with continuous twisting can be done by using Blasius equation which is used for direct channels, by considering the determinate parameters such as fluid velocity and channel length on a helical line.

Tarasov and Shchukin (1977), Vilemas and Poshkas (1992), Boltenko et al. (2001, 2007), Ustimenko (1977), Baxi and Wong (2000) and Tarasevich and Yakovlev (2010) have also worked on annular channels with a continuous twist on length.

Yakovlev et al. (2013) used swirl flow devices to enhance two-phase heat transfer. They visualized adiabatic air-water flow in horizontal transparent tubes. As a result, they observed the slug region, wave region, annular region, disperse region and cord region in the flow domain. When the liquid takes annular shape with gas flowing in the centre, then it is called annular region of the flow. As the gas content increases, some spots are found to be dry and lead to rupture in the presence of twisted tape. Thus, with the decrease in liquid content, the number of dry spots increases in the presence of twisted tape resulting in dispersing regime. Now, some of the liquid do not come into contact with the tube wall. Instead, the liquid moves around the twisted tape in the form of a cord.

This cord formation becomes significant with high gas phase content as the centrifugal forces are insignificant near the tape axis. This liquid cord formation around the tape acts as an inactive heat transfer surface. In order to prevent the cord formation, ribs can be provided on the twisted tape at an angle. This makes the heat transfer fluid, which moves along the tape to be displaced to a heat transfer surface. They concluded that the performance of twisted tape with ribs provided against the twisting direction was noted to be the most effective combination for heat transfer enhancement. More information on similar work can be gathered from Kedzierski and Kim (1997), Shchukin (1980), Tarasevich et al. (2011) and Yakovlev et al. (2011).

Omeroglu et al. (2013) studied the effect of inserts on the flow instabilities caused in boiling phenomenon. They used ring-type inserts, twisted tapes and wire coil inserts. They observed that the pressure drop was more in case of ring inserts rather than that in the case of twisted tapes. Also, the increase in pressure drop was further enhanced with the increase in pitch ratio. The use of wire coil inserts resulted in stable flow over that of twisted tapes.

Valmiki (2017) worked on a helical wire coil made up of copper and aluminium coil and varied the pitches from 1 cm, 2 cm and 3 cm. They evaluated the heat transfer coefficient of plain tube and wired coil tube and presented it in the Fig. 3.22. They concluded that Nusselt number was higher in copper material coiled wire insert

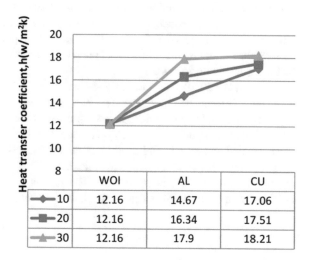

Fig. 3.22 Comparison of plain tube and insert material with heat transfer coefficient at various pitches (Valmiki 2017)

Heat transfer coefficient,h(w/m²k)

	WOI	AL	CU
10	12.16	14.67	17.06
20	12.16	16.34	17.51
30	12.16	17.9	18.21

than that of aluminium coil wire insert, and the enhancement in heat transfer was up to 1.36 and 1.28, respectively, in comparison to plain tube. They observed that wire coil performed significantly better in transition and turbulent flow conditions, and it increased heat transfer rate by 150% at a constant pumping power.

Liu and Sakr (2013) thoroughly reviewed the numerical and experimental works on twisted tape, wire coil, swirl flow generator, transverse rib, helical screw tape, coiled wire turbulators, etc. They reviewed the research work since 2004. Dewan et al. (2004) reviewed thoroughly the progress of heat transfer enhancement by using inserts, screw tape, ribs, etc. until 2004. Liu and Sakr (2013) presented Table 3.4 in which the configuration and type of inserts are described. They reported another Table 3.5 in which the working fluid, the configuration of insert, the testing conditions and their observations and conclusions are mentioned for twisted tapes.

Various wire coil insert configurations are presented in Table 3.6, and the experimental works on wire coils are tabulated in Table 3.7 with their respective thermo-hydraulic performance. Similarly, they tabulated different swirl generator configurations in Table 3.8 and thermo-hydraulic performance of swirl generator in Table 3.9. Also, they made Table 3.10 for the insert rib configurations and Table 3.11 for working fluids, operating conditions with observations and conclusions related to rib inserts. Liu and Sakr (2013) considered different swirl conical ring and their configurations in Table 3.12. They tabulated the thermo-hydraulic performance of the conical rings in Table 3.13. Overall, Liu and Sakr (2013) covered a large extent of works related to inserts that were advantageous in heat transfer enhancement.

Muñoz-Esparza and Sanmiguel-Rojas (2011) used CFD simulation for the thermo-hydraulic performance in a round pipe in which helical wire coils are inserted. They numerically simulated the incompressible laminar flow into the tube by using finite element method. They validated their numerical results with the experimental one and observed a good agreement when the Reynolds number was below 500. They observed that numerical simulation showed the trend followed by experimental results which was that friction factor became constant for Reynolds number in the range of 600–850. They simulated pitch-periodic computational domain for evaluating the influence of pitch on friction factor and found that friction factor reduced with increased pitch.

Gunes et al. (2010) experimentally studied the effect of triangular cross-sectioned wire coil inserts on heat transfer and pressure drop characteristics. They experimented under uniform heat flux boundary condition with air as a working fluid. They varied pitch ratios ($P/D = 1, 2$ and 3) in the range of 3500–27,000. They experimented and found significant increment in Nusselt number with the enhancement in Reynolds number and wire thickness. They found that Nusselt number increased with decrease in pitch ratio. The results showed that at high Reynolds number, thin wire with large coil pitch was not suitable for heat transfer

Table 3.4 Configurations of various twisted tapes (Liu and Sakr 2013)

Configuration	Name	Configuration	Name	Reference
	Typical twisted tape		Twisted tape with various widths	
	Twisted tape with various twist ratios		Short-length twisted tape	
	Left and right twist tape with rod and spacer		Peripherally cut twisted tape with an alternate axis	
	Left–right twisted tapes		Twist tape with alternate axis	
	Dual twisted tapes		Multiple twisted tape	
	Left and right twisted tape with varying space		Twisted tape with centre wing	
	Delta-winglet twisted tape		Twisted tapes with alternate axes and triangular, rectangular and trapezoidal wings	
	Twisted tape with trapezoidal cut		Serrated twisted tape at various serration depth ratios	
	Peripherally cut twisted tape		V-cut twisted tape insert	
	Jagged twisted tape		Butterfly inserts	
	Helical screw tape with varying spacers		Helical screw tape with various spacer length	
	Right and left helical screw tape		Twisted tape with oblique teeth	
	Dimpled tube fitted with twisted tape		Twisted tape consisting wire nails	

Table 3.5 Experimental works on the thermo hydraulic performance of twisted tape enhancement (Liu and Sakr 2013)

Authors	Working fluid	Configuration	Conditions	Observation
Kumar and Prasad [5]	Water	Typical twisted tape (TT)	$4000 \leq Re \leq 21000$, Twist ration $y = 3.0$–12.0	• Decreasing values of the twist-pitch to tube diameter ratio lead to increasing values of heat transfer rate, and the pressure drop as well • Twisted-tapes generate turbulency superimposed with swirlness inside the flow tube and consequently result in enhanced heat transfer • Increase in twisted-tape solar water heaters performance is remarkable at low and moderate values of the flow Reynolds number and monotonous at high values of the Reynolds number
Noothong et al. [6]	Water	Typical twisted tape (TT)	Twist ration $y = 5.0$ and 7.0 $2000 \leq Re \leq 12000$	• Swirl flow helps decrease the boundary layer thickness of the hot air flow and increase residence time of hot air in the inner tube • The enhancement efficiency and Nusselt number increases with decreasing the twist ratio and friction factor also increases with decreasing the twist ratio • Secondary fluid motion is generated by the tape twist, and the resulting twist mixing improves the convection heat transfer
Sarada et al. [7]	Air	Typical twisted tape (TT)	Width of the twisted tapes range from 10 to 22 mm $6000 \leq Re \leq 13500$	• The enhancement of heat transfer with twisted tape inserts as compared to plain tube • This enhancement is mainly due to the centrifugal forces resulting from the spiral motion of the fluid
S. Eiamsa-ard et al. [8]	Air	Full and short length twisted tape.	$4000 \leq Re \leq 20000$	• The presence of the tube with short-length twisted tape insert yields higher heat transfer rate
Jaisankar et al. [9]	Water	Full length left–right twist, fitted with rod and spacer at the trailing edge for lengths of 100, 200 and 300 mm	Twist ration $y = 3.0$ and 5.0 $700 \leq Re \leq 1600$	• The heat enhancement in full length twisted tape is better than the twist fitted with rod and spacer. The decrease in heat transfer augmentation in twist fitted with rod is minimum compared to twist fitted with spacer. The decrease in friction factor is higher for twist fitted with spacer compared to twist fitted with rod
Eiamsa-ard et al. [10]	Water	Twisted tape with wings alone (WT). Twisted tape with alternate axes alone (T-A). Typical twisted tape (TT)	Twist ratio $y = 3.0$. Angles of attack $\beta = 43°$, $53°$ and $74°$ $5200 \leq Re \leq 22000$	• WT-A with the largest angle of attack gave the highest Nusselt number (Nu), friction factor (f) as well as thermal performance factor
Wongcharee and Eiamsa-ard [11]	Water	Alternate clockwise and counter-clockwise twisted-tapes (TA)	$830 \leq Re \leq 1990$, Twist ration $y = 3.0$, 4.0 and 5.0	• The friction factor associated by TA is higher than that induced by TT, and friction factor increases with decreasing twist ratio
Eiamsa-ard et al. [12]	Air	Dual twisted tapes	$4000 \leq Re \leq 19000$, Twist ration $y = 3.0$, 4.0 and 5.0.	• The smaller space ratio of the dual twisted tapes in tandem is more attractive in heat transfer application due to higher enhancement efficiency than the single one
Eiamsa-ard [13]	Air	Multiple twisted tape vortex generators (MT-VG).	$2700 \leq Re \leq 9000$.	• The decreases of both free-spacing ratio (s/w) and twist ratio (y/w) results in the increases of Nusselt number, friction factor and also enhancement index
Jaisankar et al. [14]	Water	Twist fitted with rod	Twist ration $y = 3.0$ and 5.0	• The heat enhancement is always higher in twisted tape
Jaisankar [17]	Water	Helical and left–right twisted tapes.	Twist ratio $y = 3.0$	• The heat enhancement in helical and left–right twisted tape collectors is found to be better than the plain tube collector. While comparing the left–right and helical twisted tape collector for the same twist ratio 3, higher heat transfer and thermal performance are obtained in left–right twisted tape collector
Murugesan et al. [21]	Water	Trapezoidal-cut	$2000 \leq Re \leq 12000$, Twist ration $y = 4.4$ and 6.0	• The mean Nusselt number for trapezoidal-cut twisted tape higher than typical twisted tape
Eiamsa-ard et al. [19]	Water	Oblique delta-winglet twisted tape (O-DWT) and straight delta-winglet twisted tape	$3000 \leq Re \leq 27000$, Twist ration $y = 3.0$, 4.0 and 5.0. Depth of wing cut ratios	• The values of Nusselt number and friction factor in the test tube equipped with delta-winglet twisted tape are noticeably higher than those in the plain tube and also tube equipped with typical twisted tape

Table 3.5 (continued)

Authors	Working fluid	Configuration	Conditions	Observation
		(S-DWT), Uniform wall heat flux tube	(DR=d/w=0.11, 0.21 and 0.32)	• Nusselt number and friction factor increase with decreasing of twist ratio and increasing depth of wing cut ratio (DR) for all Reynolds numbers studied • O-DWT gives higher Nusselt number and friction factor than that of the S-DWT. The thermal performance factor in the tube with O-DWT is greater than that with S-DWT and the factor increases with decreasing Reynolds number and increasing twist ratio • DWT performs better heat transfer enhancement than that typical twisted tape • DWT can be replaced any of the TT efficiently to reduce the size of the heat exchanger
Wongcharee and Eiamsa-ard [20]	Water	Twisted tapes with wing shape including triangle, rectangle and trapezoid	5500 ≤ Re ≤ 20200, Wing-chord ratios (d/W) of 0.1, 0.2 and 0.3, Twist ratio (y/W) of 4.0.	• The twisted tapes consisted of both alternate-axes and wings offer superior heat transfer enhancement compared to the one with only alternate-axes and also the typical one. This is due to the combined effects of the strong collision of fluid behind the alternate point, caused by alternate axis and the extra fluid disturbance near tube wall induced by wings
Seemawute and Eiamsa-ard [22]	Water	Peripherally-cut twisted tape with alternate axis (PT-A)	5000 ≤ Re ≤ 20000 uniform heat flux circular tube	• Thermal performance in a tube fitted with PT-A are consistently higher than those in the tube equipped with PT, TT and also in the plain tube
Eiamsa-ard et al. [23]	Water	Peripherally-cut twisted tape	1000 ≤ Re ≤ 20000, Twist ratio y=3.0	• The peripherally-cut twisted tape offered higher heat transfer rate, friction factor and also thermal performance factor compared to the typical twisted tape. An additional turbulence of fluid in the vicinity of the tube wall and vorticity behind the cuts generated by the modified twisted tape are referred as the reason for a better heat transfer enhancement
Eiamsa-ard and Promvonge [30]	Air	Serrated twisted tape (STT)	4000 ≤ Re ≤ 20000, Twist ratio y=4.0	• The STT gives higher heat transfer rate than the TT while yields higher friction factor than the TT
Murugesan et al. [31]	Water	Twisted tape consisting wire nails (WN-TT)	2000 ≤ Re ≤ 12000, Twist ratios y=2.0, 4.4 and 6.0	• The better performance of WN-TT is due to combined effects of the following factors: 1) common swirling flow generated by P-TT 2) additional turbulence offered by the wire nails
Jaisankar et al. [20]	Water	Helical twisted tape	Twist ratios y=3.0, 4.0, 5.0 and 6.0	• Thermal performance of twisted tape collector with minimum twist ratio (Y=3) is better than the other twist ratios
Ibrahim [23]	Water	Helical screw element	570 ≤ Re ≤ 1310 y=2.17, 3.33, 4.3, and 5	• The averaged Nusselt number, Nu increase with the increase in Reynolds number and with the decrease in twist ratio and spacer length
Murugesan et al. [24]	Water	V-cut twisted tape	2000 ≤ Re ≤ 12000	• The V-cut twisted tape offered a higher heat transfer rate, friction factor and also thermal performance factor compared to the plain twisted tape. In addition, the influence of the depth ratio was more dominant than that of the width ratio for all the Reynolds number • The thermal performance factors for all the cases are more than one indicating that the effect of heat transfer enhancement due to the enhancing tool is more dominant than the effect of the rising friction factor and vice versa • Nusselt number and the mean friction factor in the tube with V-cut twisted tape (VTT) increase with decreasing twist ratios (y), width ratios (WR) and increasing depth ratios (DR)
Moawed [25]	Water	Helical screw element	570 ≤ Re ≤ 1310	• The averaged Nusselt number Nu increases with an increase in the Reynolds number and with a decrease in Y and S • The Nu of the plain elliptic tube is greater than that of the plain circular tube and the Nu of elliptic tubes containing a helical screw tapes is better than that of the plain elliptic tubes for all Re, Y, and S

Table 3.5 (continued)

Authors	Working fluid	Configuration	Conditions	Observation
Sivashanmugam and Nagarajan [27]	Water	Right–left helical screw inserts of equal length	$200 \le Re \le 3000$ $y = 2.93$–4.89	• The heat transfer coefficient enhancement for right–left helical screw inserts is higher than that for straight helical twist for a given twist ratio
Sivashanmugam and Suresh [28]	Water	Full-length helical screw element of different twist ratio	$v = [0.1 \times 10^{-3}$ to $2.4 \times 10^{-3}]\,m^3/min$	• There is no much change in the magnitude of heat transfer coefficient enhancement with decreasing twist ratio and with increasing twist ratio, as the intensity of swirl generated at the inlet or at the outlet in the order of increasing twist ratio or decreasing twist ratio, is same in both the cases
Krishna et al. [29]	Water	Straight full twist	Twist ratio $y = 4.0$	• The heat transfer coefficient increases with Reynolds number and decreasing spacer distance with maximum being 2 in. spacer distance for both the type of twist inserts. Also, there is no appreciable increase in heat transfer enhancement in straight full twist insert with 2 in. spacer distance
Thianpong et al. [32]	Water	Dimpled tube fitted with a twisted tape swirl generator	$12000 \le Re \le 44000$, Twist ratios $y = 3.0, 5.0$, and 7.0	• A dimpled tube in common with a twisted tape has significant effects on the heat transfer enhancement and friction factor. The heat transfer and friction factor are increase with decreasing both of pitch ratio (PR) and twist ratio (y/w)
Saha [33]	air	Twisted-tape inserts with and without oblique teeth	$10000 \le Re \le 100000$	• Full-length and short-length twisted-tapes with oblique teeth in combination with axial corrugations show only marginal improvements over the twisted-tapes without oblique teeth

Table 3.6 Different wire coil inserts (Liu and Sakr 2013)

Configuration	Name	Configuration	Name
	Coiled square wires		Twisted tape and wire coil
	Non-uniform wire coil combined with twisted tape		Triangle cross-sectioned coiled wire
	Coiled wire turbulators		Wire coil in pipe

enhancement. They calculated that best enhancement efficiency was 36.5 obtained at Reynolds number 3858. This maximum enhancement efficiency was obtained at distinct wire width $a/D = 0.0892$ and pitch ratio $P/D = 1$.

Table 3.7 Experimental works related to the thermo-hydraulic performance of wire coil enhancement (Liu and Sakr 2013)

Authors	Walking fluid	Configuration	Conditions	Observation
Gararcia et al. [38]	Water–propylene glycol mixtures	• $1.17 \leq p/d \leq 2.68$ (helical pitch) • $0.07 \leq e/d \leq 0.10$ (wire diameter)	• $80 \leq Re \leq 90{,}000$ • $2\,8 \leq Pr \leq 150$	• In laminar flow, results show that wire coils behave mainly as a smooth tube • In turbulent flow, wire coils cause a high pressure drop increase which depends mainly on pitch to wire diameter ratio (p/e)
Yakut and Sahin [39]	Air	• Coiled wire cross section 4 mm • Coiled wire length 1240 mm	• $5000 \leq Re \leq 35{,}000$ • Pitches (10, 20, 30 mm)	• Vortex characteristics of the turbulators should be considered as a selecting criterion with heat transfer and friction characteristics in heat transfer enhancement applications
Promvonge [40]	Air	• Wires with square cross section	• $5000 \leq Re \leq 25{,}000$	• The Nusselt number augmentation tends to decrease rapidly with the rise of Reynolds number • The coiled square wire should be applied instead of the round one to obtain higher heat transfer and performance, leading to more compact heat exchanger
Promvonge [41]	Air	• Coiled wires in conjunction with a snail-type swirl generator	• $5000 \leq Re \leq 25{,}000$	• The use of the coiled wires and the snail entry causes a high pressure drop
Promvonge [42]	Air	• Coil wires in conjunction with twisted tapes	• $3000 \leq Re \leq 18{,}000$	• Nusselt number augmentation tends to decrease with the rise of Reynolds number • The combined wire coil and twisted tape turbulators are compared with a smooth tube at a constant pumping power. Heat transfer performance is doubled at low Reynolds number

(continued)

Table 3.7 (continued)

Authors	Walking fluid	Configuration	Conditions	Observation
				• The best operating regime for combined turbulators is found at lower Reynolds number values for the lowest values of the coil spring pitch and twist ratio
Eiamsa-ard etal. [43]	Air	• Combined devices of the twisted tape (TT) and constant/periodically varying wire coil pitch ratio	• $4600 \leq Re \leq 20{,}000$	• At low Reynolds number, the compound devices of the TT with $Y = 3$ and the DI-coil provide the highest thermal performance
Gunes [44]	Air	• Pitch ratios $P/D = 1$, 2 and 3 • $a/D = 0.0714$ and 0.0892	• $3500 \leq Re \leq 27{,}000$	• The equilateral triangle cross-sectioned coiled wire inserted separately from the tube wall and the coiled wire inserts induced a remarkable increase in both pressure drop and heat transfer in comparison with the smooth tube depending on coil pitches and wire thickness
Akhavan-Behabadi et al [45]	Oil	• Seven coiled wires having pitches of 12–69 mm	• $10 \leq Re \leq 1500$	• Wire coil inserts with small wire diameters have better performance, especially at low Reynolds numbers. Also, the increase in the coil pitch made a moderate decrease in performance parameter

Table 3.8 Different swirl generators (Liu and Sakr 2013)

Configuration	Name	Configuration	Name
	Conical injector		Propeller swirl generators
	Radial guide vane swirl generators		Typical swirler
	Propeller swirl generator		

Table 3.9 Experimental works related to the thermo-hydraulic performance of swirl generators enhancement (Liu and Sakr 2013)

Authors	Working fluid	Configuration	Conditions	Observation
Yilmaz et al. [48]	Air	• The swirl generator with conical deflecting element • The swirl generator with spherical deflecting element	• $32,000 \leq Re \leq 111,000$	• The swirl generator with no deflecting element may be advantageous in terms of heat transfer enhancement and energy saving in comparison with swirl generators with a deflecting element • In swirling flow, increasing the Reynolds number and the vane angle increased the Nusselt number • To obtain lower pumping powers for the same heat transfer rate, higher vane angles and relatively lower Reynolds numbers must be employed
Saraç and Bali [49]	Air	• Propeller-type geometry	• $5000 \leq Re \leq 30,000$ • Positions of the vortex generator in the axial direction are examined: at the inlet $x = 0$, $x = L/4$ and $x = L/2$	• For the decaying swirl flow, the heat transfer and pressure drop decreased with the axial distance • The inserts with six vanes resulted in more heat transfer values than those with four vanes

(continued)

Table 3.9 (continued)

Authors	Working fluid	Configuration	Conditions	Observation
				• For the decaying swirl flow, the heat transfer and pressure drop decreased with the axial distance • The inserts with six vanes resulted in more heat transfer values than those with four vanes
Kurtbs et al. [50]	Air	• Conical injector-type swirl generator (CITSG)	• $9400 \leq Re \leq 35{,}000$ • CITSG angle (α) of 30°, 45° and 60°	• The heat transfer ratio (Nu_{PR}) decreases with increase in Re • The effect of α on Nu_x is at a negligible level for higher Re • The heat transfer enhancement ratio (Nu_{ER}) decreases with the increase in Re and increases with the increase in β and DR
Eiamsa-ard et al [51]	Air	• Propeller-type swirl generators at several pitch ratios (PR)	• $4000 \leq Re \leq 21{,}000$ • Blade numbers of the propeller ($N = 4$, 6 and 8 blades) • Different blade angles ($\theta = 30°$, 45°, and 60°)	• The heat transfer in the test tube can be enhanced considerably by insertion of the propeller-type swirl generators. The heat transfer and the enhancement efficiency are found to increase with increasing the blade number (N) and the blade angle (θ) but to decrease with the rise of pitch ratio (PR)
Yang et al. [52]	Air	• Flat vane swirler situated at the entrance of the pipe	• $7970 \leq Re \leq 47{,}820$	• The Nusselt number is found to increase monotonically with both the Reynolds and the swirl numbers in the weak swirling flow region • The Nusselt number decreases with increase in the swirl numbers due to suppression effect of accelerated circumferential flow

Table 3.10 Different insert ribs (Liu and Sakr 2013)

Configuration	Name	Configuration	Name
	Inclined ribs		Discrete double-inclined ribs
	Helical rib		Combined rib with rectangular winglet
	Combined wedge ribs and winglet-type vortex	Pointing Downstream, PD Pointing Upstream, PU	Combined rib with delta-winglet
In-line staggered	Triangular rib		Louvred strip

Table 3.11 Experimental works on the thermo-hydraulic performance of insert rib enhancement (Liu and Sakr 2013)

Authors	Working fluid	Configuration	Conditions	Observation
Naphon et al.	Water	• Horizontal double-pipes with helical ribs	• $15,000 \leq Re \leq 60,000$	• Helical ribs have a significant effect on the heat transfer and pressure drop augmentations
Li et al.	Water	• Discrete double-inclined ribs	• $15,000 \leq Re \leq 60,000$	• Visualization of the flow field shows that in addition to the front and rear vortices around the ribs, main vortices and induced vortices are also generated by the ribs in the DDIR tube. The rear vortex and the main vortex contribute much to the heat transfer enhancement in the DDIR tubes
Eiamsa et al.	Water	• Louvred strip	• $6000 \leq Re \leq 42,000$ • Forward or backward arrangements • nclined angles ($\theta = 15°, 25°$ and $30°$)	• Louvred strip insertions can be used efficiently to augment heat transfer rate because of the turbulence intensity • The highest heat transfer rate was

(continued)

Table 3.11 (continued)

Authors	Working fluid	Configuration	Conditions	Observation
				achieved for the backward inclined angle of 30° due to the increase of strong turbulence intensity
Depaiwa et al.	Air	• Rectangular winglet vortex generator (WVG)	• $5000 \leq Re \leq 23,000$ • Attack angles (α) of 60°, 45° and 30° • $DW = b/H = 0.4$ • Transverse pitch, $P_t/H = l$	• Solar air heater channel with rectangular WVG provides significantly higher heat transfer rate and friction loss than the smooth wall channel
Chompookham et al.	Air	• Combined wedge ribs and winglet-type vortex generators (WVGs)	• $5000 \leq Re \leq 22,000$ • Uniform heat-flux tube • Attack angle 60° • Rib height, $e/H = 0.2$ • Rib pitch, $P/H = 1.33$	• The combination of staggered wedge rib and the WVGs has efficiently performed and should be applied to obtain higher thermal performance of about 17–20% of a single use of turbulators
Promvonge et al.	Air	• Combined ribs and delta-winglet-type vortex generators (DWs)	• $5000 \leq Re \leq 22,000$ • Attack angles (α) of 60°, 45° and 30°	• The Nu decreases slightly with the rise in Re • Combined rib and PD-DW at lower angle of attack provides higher heat transfer • The best operating regime for using these compound turbulators is found at the PD-DW arrangement, lower attack angle and/or Re values
Promvonge et al.	Air	• Combined wedge ribs and winglet-type vortex generators (WVGs)	• $5000 \leq Re \leq 22,000$ • Attack angles 30°, 45°, 60°	• The Nusselt number augmentation tends to decrease slightly with the rise in Reynolds number • The best operating regime for using these compound turbulators is found at the lower attack angle and/or Reynolds number values

Table 3.12 Different conical swirl rings (Liu and Sakr 2013)

Configuration	Name	Configuration	Name
	Converging-diverging conical ring		Diverging conical ring
	Converging conical ring		Conical ring with twisted tape
	Conical nozzle combined with a snail		Conical nozzle
	Conical nozzle tabulator		V-nozzle turbulator
	V-nozzle turbulator		Tandem diamond-shaped tabulator
	Conical tube insert		Perforated conical rings

Table 3.13 Experimental works on the thermo-hydraulic performance of conical ring enhancement (Liu and Sakr 2013)

Authors	Working fluid	Configuration	Conditions	Observation
Promvonge	Air	• Conical ring arrangements	• $6000 \le Re \le 26{,}000$	• Although the effect of using the conical ring causes a substantial increase in friction factor
Promvonge and Eiamsa-ard	Air	• Conical ring turbulators and a twisted tape swirl generator	• $6000 \le Re \le 26{,}000$	• For all the devices used, the enhancement efficiency tends to decrease with the rise of Reynolds number and to be nearly uniform for Reynolds over 16,000 • The smaller the twist ratio is, the larger the heat transfer and friction for all Reynolds numbers
Promvonge and Eiamsa-ard	Air	• Conical nozzles and swirl generator	• $8000 \le Re \le 18{,}000$ • Pitch ratios (PR) of conical nozzle arrangements, PR = 2.0,	• The heat transfer in the circular tube could be enhanced considerably by fitting it with conical nozzle inserts and small entrance. Although

(continued)

Table 3.13 (continued)

Authors	Working fluid	Configuration	Conditions	Observation
			4.0 and 7.0 • Uniform heat flux	they provide higher energy loss of the fluid flow, the loss is low especially at low Reynolds number • The turbulators are applicable effectively at low Reynolds number because they provide low enhancement efficiency at high Reynolds number for all PRs
Promvonge and Eiamsa-ard	Air	• Conical nozzle turbulators	• $8000 \leq Re \leq 18{,}000$ • Uniform heat-flux tube • Conical or converging nozzle (C-nozzle) • $PR = 2.0, 4.0$ and 7.0	• C-nozzle turbulators and a small free-spacing entry can be employed effectively at low Reynolds number or in places where pumping power is not important, but compact sizes and ease of manufacture are needed
Promvonge and Eiamsa-ard	Air	• Conical-nozzle turbulator	• $8000 \leq Re \leq 18{,}000$ • $PR = 2.0, 4.0$ and 7.0 • Diverging nozzle arrangement (D-nozzle) • Converging nozzle arrangement (C-nozzle)	• The heat transfer rate in the tube can be promoted by fitting with nozzle turbulators. Despite very high friction, the turbulators can be applied effectively in places where pumping power is not significantly taken into account but the compact size including ease of manufacture installation is required
Promvonge and Eiamsa-ard	Air	• V-nozzle turbulator inserts in conjunction with a small entry	• $8000 \leq Re \leq 18{,}000$ • $PR = 2.0, 4.0$ and 7.0	• V-nozzle alone provides the best thermal performance over other turbulator devices
Eiamsa-ard and Promvonge	Air	• V-nozzle turbulator inserts	• $8000 \leq Re \leq 18{,}000$ • $PR = 2.0, 4.0$ and 7.0	• The enhancement efficiency decreases with increasing Reynolds number
Eiamsa-ard and Promvonge	Air	• Tandem diamond-shaped	• $3500 \leq Re \leq 16{,}500$ • Cone angle	• The increase in the heat transfer rate by increasing the cone

(continued)

Table 3.13 (continued)

Authors	Working fluid	Configuration	Conditions	Observation
		turbulators (D-shape turbulator)	$(\theta = 15°, 30°$ and $45°)$ • Tall length ratio $(TR = l_r/l_h = 1.0,$ 1.5 and 2.0)	angle and decreasing the tall length ratio is due to the higher turbulence intensity imparted to the flow between the turbulators and the heating wall. Since the turbulators are placed directly into the flow core, they cause high friction losses because of high flow blockage
Anvarl et al.	Water	• Conical tube inserts	• $2500 \leq Re \leq 9500$	• The insertion of turbulators has significant effect on the enhancement of heat transfer, especially the DR arrangement, and also they increase the pressure drop. So turbulators can be used in places where the compact size is more significant than pumping power
Kongkaltpalboon et al.	Air	• Perforated conical ring (PCR)	• $4000 \leq Re \leq 20,000$ • Pitch ratios (PR, $p/D = 4, 6$ and 12) • Perforated holes ($N = 4,$ 6 and 8 holes)	• The heat transfer rate and friction factor of PCRs increase with decreasing pitch ratio (PR) and decreasing number of perforated hole (N). However, the thermal performance factor increase with increasing number of perforated hole and decreasing pitch ratio

References

Ahire D, Bhalerao Y, Dholkawala M (2017) Experimental investigation of heat transfer enhancement in circular pipe using conical ring inserts. Int J Mech Prod Eng 5(3):148–153

Abu-Mulaweh HI (2003) Experimental comparison of heat transfer enhancement methods in heat exchangers. Int J Mech Eng Educ 31(2):160–167

Agrawal KN, Kumar R, Lal SN, Varma HK (2004) Heat transfer augmentation by segmented tape inserts during condensation of R-22 inside a horizontal tube. ASHRAE Trans 110:143–150

Akhavan-Behabadi MA, Salimpur MR, Pazouki VA (2008) Pressure drop increase of forced convective condensation inside horizontal coiled wire inserted tubes. Int Commun Heat Mass Transf 35:1220–1226

Baxi CB, Wong CPC (2000) Review of helium cooling for fusion reactor applications. Fusion Eng Des 51–52:319–324

Bergles AE (1985) Techniques to augment heat transfer (Chap. 3). In: Rohsenow WM, Hartnett JP, Ganie EN (eds) Handbook of heat transfer applications, 2nd edn. McGraw-Hill, New York

Boltenko EA, Tarasevich SE, Obuhova LA (2001) Heat transfer intensification in annular channels with a flow twisting. A convective heat transfer. Izv Ross Akad Nauk, Energetika 3:99–104. (in Russian)

Boltenko EA, Il'in GK, Tarasevich SE, Yakovlev AB (2007) Heat transfer in annular channels with flow twisting. Russ Aeronaut (Iz VUZ) 50(3):287–291

Bruening GB, Chang WS (1999) Cooled cooling air systems for turbine thermal management. In: Int gas turbine & aeroengine congress & exhibition ASME 99-GT-14

Chen L, Zhang HJ (1993) Convection heat transfer enhancement of oil in a circular tube with spiral spring inserts. In: Chow LC, Emery AF (eds) Heat transfer measurements and analysis, HTD-ASME Symp., vol 249, pp 45–50

Colburn AP, King WJ (1931) Relationship between heat transfer and pressure drop. Ind Eng Chem 23(8):918–923

Dewan A, Mahanta P, Sumithra Raju K, Suresh Kumar P (2004) Review of passive heat transfer augmentation techniques. Proc Inst Mech Eng A J Power Energy 218(7):509–527

Durgesh VA, Yogesh JB, Murtaza SD (2017) Experimental investigation of heat transfer enhancement in circular pipe using conical ring inserts. International J Mechanical Production Engineering 5(3):148–153

Edwards FJ, Sheriff N (1961) The heat transfer and friction characteristics for forced convection air flow over a particular type of rough surface. In: International developments in heat transfer. ASME, New York, pp 415–425

Emerson WH (1961) Heat transfer in a duct in regions of separated flow. In: Proceedings of the third international heat transfer conference, vol 1, pp 267–275

Evans LB, Churchill SW (1963) The effect of axial promoters on heat transfer and pressure drop inside a tube. Chem Eng Prog Symp Ser 59(41):36–46

Fan A, Deng J, Nakayama A, Liu W (2012) Parametric study on turbulent heat transfer and flow characteristics in a circular tube fitted with louvered strip inserts. Int J Heat Mass Transf 55 (19):5205–5213

Garcia A, Vicente PG, Viedma A (2005) Experimental study of heat transfer enhancement with wire coil inserts in laminar-transition-turbulent regimes at different Prandtl numbers. Int J Heat Mass Transf 48(21–22):4640–4651

Gunes S, Ozceyhan V, Buyukalaca O (2010) Heat transfer enhancement in a tube with equilateral triangle cross sectioned coiled wire inserts. Exp Thermal Fluid Sci 34:684–691

Guo J, Fan A, Zhang X, Liu W (2011) A numerical study on heat transfer and friction factor characteristics of laminar flow in a circular tube fitted with center-cleared twisted tape. Int J Therm Sci 50(7):1263–1270

Hejaji V, Akhavan-Behabadi MA, Afshari A (2010) Experimental investigation of twisted tape inserts performance on condensation heat transfer enhancement and pressure drop. Int Commun Heat Mass Transf 37:1376–1387

Herring NR, Heister SD (2009) On the use of wire-coil inserts to augment tube heat transfer. J Enhanc Heat Transf 16(1):19–34

Hsieh S-S, Kuo M-T (1994) An experimental investigation of the augmentation of tube-side heat transfer in a crossflow heat exchanger by means of strip-type inserts. J Heat Transf 116:381–390

Inaba H, Ozaki K (1997) Heat transfer enhancement and flow-drag reduction of forced convection in circular tubes by means of wire coil insert," Proc. Int. Conf. on Compact Heat Exchangers for the Process Ind., In: Shah RK, Bell KJ, Mochizuki S, Wadekar VW (eds) Begell House Inc., New York, pp 445–452

Kedzierski MA, Kim MS (1997) Convective boiling and condensation heat transfer with a twisted-tape insert for R12, R22, R152a, R134a, R290, R32/R134a, R32/R152a, R290/R134a, R134a/R600a, Report NISTIR 5905. National Institute of Standards and Technology, Gaithersburg, MD

Khatua AK, Kumar P, Singh HN (2014) Pressure drop in R-245fa condensation in tubes fitted with coiled-wire inserts. J Enhanc Heat Transf 21(6):425–438

Koch R (1958) Druckverlust und Waerrneuebergang bei verwirbeiter Stroemung, Vei: Dtsch. lngen. Forschungsheft, Ser. B 24(469):1–44

Liu S, Sakr M (2013) A comprehensive review on passive heat transfer enhancements in pipe exchangers. Renew Sust Energ Rev 19:64–81

Liu X, Li C, Cao X, Yan C, Ding M (2018) Numerical analysis on enhanced performance of new coaxial cross twisted tapes for laminar convective heat transfer. Int J Heat Mass Transf 121:1125–1136

Manglik RM, Bergles AE (1992a) Heat transfer and pressure drop correlations for twisted-tape inserts in isothermal tubes: part I—laminar flows. Enhanc Heat Transf ASME Symp HTD 202:89–98

Manglik RM, Bergles AE (1992b) Heat transfer and pressure drop correlations for twisted-tape inserts in isothermal tubes: part II—transition and turbulent flows. Enhanc Heat Transf ASME Symp HTD 202:99–106

Manglik RM, Bergles AE (1993a) Heat transfer and pressure drop correlations for twisted-tape inserts in isothermal tubes: part I—laminar flows. J Heat Transf 115:9

Manglik RM, Bergles AE (1993b) Heat transfer and pressure drop correlations for twisted-tape inserts in isothermal tubes: part II—transition and turbulent flows. J Heat Transf 115 (4):890–896

Megerlin FE, Murphy RW, Bergles AE (1974) Augmentation of heat transfer in tubes by means of mesh and brush inserts. J Heat Transf 96:145–151

Muñoz-Esparza D, Sanmiguel-Rojas E (2011) Numerical simulations of the laminar flow in pipes with wire coil inserts. Comput Fluids 44(1):169–177

Naphon P (2006) Heat transfer and pressure drop in the horizontal double pipes with and without twisted tape insert. Int Commun Heat Mass Transf 33:166–175

Oliver DR, Aldington RWJ (1986) Enhancement of laminar flow heat transfer using wire matrix turbulators. In: Heat transfer—1986, proceedings of the eighth international heat transfer conference, vol 6, pp 2897–2902

Omeroglu G, Gomakh O, Karagoz S, Karsli S (2013) Comparison of the effects of different types of tube inserts on two-phase flow instabilities. J Enhanc Heat Transf 20(2):179–194

Ould Didi MB, Kattan N, Thome JR (2002) Prediction of two-phase pressure gradients of refrigerants in horizontal tubes. Int J Refrig 25:935–947

Ozceyhan V, Gunes S, Buyukalaca O, Altuntop N (2008) Heat transfer enhancement in a tube using circular cross sectional rings separated from wall. Appl Energy 85(10):988–1001

Park Y, Cha J, Kim M (2000) Heat transfer augmentation characteristics of various inserts in a heat exchanger tube. J Enhanc Heat Transf 7:23–34

Saha S, Gaitonde UN, Date A (1989) Heat transfer and pressure drop characteristics of laminar flow in a circular tube fitted with regularly spaced twisted-tape elements. Exp Thermal Fluid Sci 2 (3):310–322

Said SA, Azer NZ (1983) Heat transfer and pressure drop during condensation inside horizontal tubes with twisted tape inserts. ASHRAE Trans 89:96–113

Saysroy A, Eiamsa-ard S (2017) Periodically fully-developed heat and fluid flow behaviours in a turbulent tube flow with square-cut twisted tape inserts. Appl Therm Eng 112:895–910

Sethumadhavan R, Raja Rao M (1983) Turbulent flow heat transfer and fluid friction in helical wire coil inserted tubes. Int J Heat Mass Transf 26:1833–1845

Shchukin VK (1980) Heat transfer and hydrodynamics of internal flows in a mass-force field. Mashinostroenie in Russian, Moscow

Sivashanmugam P, Suresh S (2006) Experimental studies on heat transfer and friction factor characteristics of laminar flow through a circular tube fitted with helical screw-tape inserts. Appl Therm Eng 26(16):1990–1997

Tarasevich SE, Yakovlev AB (2010) Heat transfer in annular channel with continuous flow twisting. In: In Proc. of international heat transfer conference, Washington, DC, paper IHTC14-22617, pp 609–617

Tarasevich SE, Yakovlev AB, Giniatullin AA, Shishkin AV (2011) Heat and mass transfer in tubes with various twisted tape inserts. In: Proc. of the ASME international mechanical engineering congress & exposition, IMECE2011, Denver, CO, USA, Paper IMECE2011–62088, pp 1–6

Tarasov GI, Shchukin VK (1977) An experimental study of heat transfer in channels equipped with extended screwtype intensifiers. Heat Mass Transf Aircraft Engines, Kazan Aviation Institute, Kazan, Russia 1:40–45

Tu W, Tang Y, Hu J, Wang Q, Lu L (2015) Heat transfer and friction characteristics of laminar flow through a circular tube with small pipe inserts. Int J Therm Sci 96:94–101

Ustimenko BP (1977) Processes of turbulent carrying over in twirled currents. Nauka, Alma-Ata, USSR

Uttarwar SB, Raja Rao M (1985) Augmentation of laminar flow beat transfer in tubes by means of wire coil inserts. J Heat Transf 105:930–935

Valmiki S (2017) Heat transfer enhancement in pipe with passive enhancement technique. Int J Engg Dev Tech 5(3)

Vilemas Y, Poshkas P (1992) Heat transfer in gas-cooled channels under the effect of thermal-gravity and centrifugal forces. Academia, Vilnius, Lithuania

Wen MY, Jang KJ, Ho CY (2015) Flow boiling heat transfer in R-600a flows inside an annular tube with metallic porous inserts. J Enhanc Heat Transf 22(1):47–65

Xie L, Gu R, Zhang X (1992) A study of the optimum inserts for enhancing convective heat transfer of high viscosity fluid in a tube. In: Chen X-J, Vezirogln TN, Tien CL (eds) Multiphase flow and heat transfer; Second international symposium, vol 1. Hemisphere, New York, pp 649–656

Yakovlev AB (2013) Heat transfer and hydraulic resistance in single-phase forced convection in annular channels with twisting wire inserts. J Enhanc Heat Transf 20(6):519–525

Yakovlev AB, Tarasevich SE, Ilyin GK, Shchelchkov AV (2011) The device for a heat exchange intensification in channels of various cross-section section. Patent for the invention RU No. 2432542 C2, Demand No. 2009147927

Yakovlev AB, Tarasevich SE, Giniyatullin AA, Shishkin AV (2013) Hydrodynamics and heat transfer in tubes with smooth and ribbed twisted tape inserts. J Enhanc Heat Transf 20 (6):511–518

Zhang X, Liu Z, Liu W (2013) Numerical studies on heat transfer and friction factor characteristics of a tube fitted with helical screw-tape without core-rod inserts. Int J Heat Mass Transf 60:490–498

Chapter 4
Swirl Generators, Extended Surface Insert and Tangential Injection Devices

4.1 Swirl Generators

Swirl generators are useful in turbo machinery, pollution control devices, combustion chambers, fusion reactors, etc. Swirl flow is grouped into two parts: (1) continuous swirl flow and (2) decaying swirl flow. The continuous swirl flow keeps up its characteristic throughout the testing section, whereas decaying swirl flow dies out within the testing section. Some good research on this topic is presented here.

Sarac and Bali (2007) examined the swirl flow in a horizontal tube which was generated by vortex generator with propeller-type geometry. They studied the thermo-hydraulic characteristic of swirl flow and found that Nusselt number was increased from 181 to 163% depending on the variation of Reynolds number. It also depends on the number of the vanes, the angle and the position of the vortex generator. They achieved increased value of heat transfer with the inserts involving six vanes as compared to that with four vanes. They observed that heat transfer and pressure drop were inversely proportional to the distance from axis.

Eiamsa-ard et al. (2009) used propeller-type swirl generator with varying pitch ratios and studied their heat transfer and pressure drop characteristics. They concluded that the maximum enhancement efficiency increased up to 1.2 by using the propeller. It was observed that propeller blade number has a significant role in heat transfer enhancement. The friction factor was 3–18 times higher than that of plain tube. They observed that increasing blade number (N) and blade angle (θ) increased both the enhancement efficiency and heat transfer. The heat transfer depends on Reynolds number. They observed that for $Pr = 5.0$, 7.0 and 10.0, there was increase in heat transfer rate of approximately 113%, 90% and 73%, respectively above the plain tube.

Martemianov and Okulov (2004) numerically established a model for axisymmetric swirl pipe flow to study heat transfer characteristic. The study showed that inefficiency of the traditional correlation for convective heat transfer is

© The Author(s), under exclusive license to Springer Nature Switzerland AG 2020
S. K. Saha et al., *Insert Devices and Integral Roughness in Heat Transfer Enhancement*, SpringerBriefs in Applied Sciences and Technology,
https://doi.org/10.1007/978-3-030-20776-2_4

Table 4.1 Vortex parameters in swirl flow with vortex breakdown (Martemianov and Okulov 2004)

	Integral flow parameters								
	$\dfrac{G/}{RU}$	$\dfrac{Q/\rho}{UR^2}$	$\dfrac{M/\rho}{U^2R^3}$	$\dfrac{J/\rho}{U^2R^2}$	$\dfrac{E/\rho}{U^3R^2}$				
	0.37	3.1	1.0	21.4	18.6				
	Vortex parameters								
Vortex type	$\dfrac{a}{R}$	$\dfrac{\Gamma}{UR}$	$\dfrac{2\pi l}{R}$	$\dfrac{w_0}{U}$	$\dfrac{P_0}{\rho U^2}$	$\dfrac{V}{U}$	$\dfrac{W}{U}$	$\left(\dfrac{Re_W}{Re_U}\right)^{1/2}$	$\left(\dfrac{Re_{(V,W)}}{Re_U}\right)^{1/2}$
Approximation of cortex core by Gaussian distribution of the vorticity (4.2)									
Right helix	*0.17*	*0.37*	*0.27*	*2.23*	*0.00*	*0.37*	*0.86*	*0.93*	*0.96*
Left helix	0.25	0.37	−0.63	0.39	1.89	0.37	1.01	1.00	1.01
Left helix	*0.32*	*0.37*	*−0.26*	*−0.27*	*2.32*	*0.37*	*1.15*	*1.07*	*1.09*
Left helix	0.46	0.39	−0.14	−0.96	2.32	0.39	1.76	1.32	1.34
Approximation of cortex core by rational distribution of the vorticity (4.3)									
Right helix	0.11	0.38	0.29	2.13	−2.79	0.37	0.84	0.92	0.95
Left helix	0.31	0.43	−0.15	−1.02	2.14	0.39	1.54	1.24	1.26
Left helix	0.48	0.51	−0.12	−1.23	2.18	0.41	2.12	1.46	1.46

$$Nu \cdot f(Pr) = aRe_U^b \tag{4.1}$$

They proposed the correlation and described as

$$Nu/Nu_o = \left(\frac{Re_W}{Re_U}\right)^{1/2} = \left(W/U\right)^{1/2} \tag{4.2}$$

$$Nu/Nu_o = \left(\frac{Re_{(V,W)}}{Re_U}\right)^{1/2} = \left(\frac{W}{U}\sqrt[3]{\left(1 + V^2/W^2\right)}\right)^{1/2} \tag{4.3}$$

They presented Table 4.1 which shows that considering similar integral flow characteristics, the rate of heat transfer can be raised up to 54% by changing the vortex symmetry and controlled flow parameters, and it is possible to transit from one type to another type of vortex symmetry. This transition can probably enhance the heat transfer characteristic. Moawed (2011) conducted experimental studies related to the thermo-hydraulic performance of the circular and elliptical tubes in the laminar flow conditions. The tube consisted of full-length helical screw tape inserts. Figure 4.1 represented the elliptical tube and helical screw tape inserts. The objective of the study was to evaluate the impact of different twist ratios (Y) and pitch ratios (S) on friction factor and heat transfer characteristics for a range of Reynolds number 5700–13,100. They concluded the Nusselt number for circular tube with the correlation

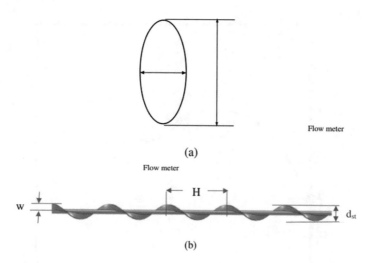

(a)

Flow meter

(b)

Fig. 4.1 (a) Cross section of the elliptic tube. (b) Configuration of the inserted tape (Moawed 2011)

$$Nu = 1.86\left(Re \cdot Pr \cdot {}^{D}\!/_{L}\right)^{0.33}\left({}^{\mu_a}\!/_{\mu}\right)_{b}^{0.14} \tag{4.4}$$

and for elliptical tube, it was computed from

$$Nu = 1.35Re^{0.21}S^{-0.183}Y^{-0.63} \tag{4.5}$$

The friction factor for elliptical tube was presented as

$$f = 40.45Re^{-0.92}S^{-0.131}Y^{-0.229} \tag{4.6}$$

They varied twist ratio from 0.22 to 0.35 and pitch ratio from 0.46 to 2.15. From the experimental analysis, they concluded that Nusselt number increased with Reynolds number. Also, Nusselt number raised by decreasing twist ratio and pitch ratio. They found that plain elliptical tube performed better as compared to plain circular tube. They found that elliptical tube arranged with helical screw tapes performed better than that of plain elliptical tube, and it was valid for all twist ratio, pitch ratio and Reynolds number taken into consideration. They observed that performance parameter (η) increased with increasing S and Y up to $S = 1$ and $Y = 0.22$, and then it degraded. These results are presented in Figs. 4.2 and 4.3 respectively.

Hussain et al. (2016) studied the effect of a pair of winglet vortex generators on the heat transfer from an endwall of a protruded circular cylinder. Two orientations of the winglet vortex generators that is counterflow inward (CFI) and counterflow outward (CFO) have been considered for the study on the upstream of the cylinder.

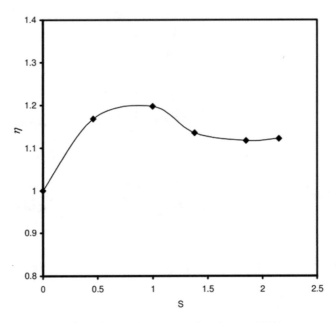

Fig. 4.2 Variation of η with S at $Y = 0.31$ and $Re = 1279$ (Moawed 2011)

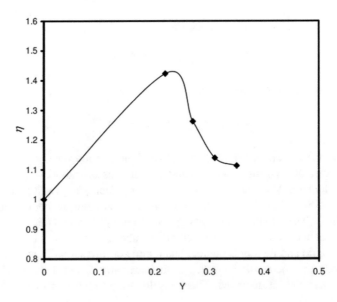

Fig. 4.3 Variation of η with Y at $S = 1$ and $Re = 1279$

The winglet vortex generator pair has been shown in Fig. 4.4. The schematic representation of the counterflow inward and counterflow outward orientations of the winglet vortex generator pair has been shown in Fig. 4.5. They observed that heat

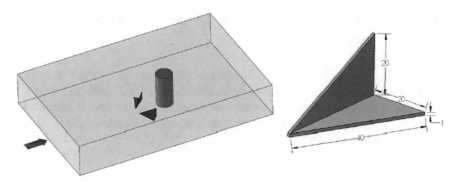

Fig. 4.4 Winglet vortex generator pair (Hussain et al. 2016)

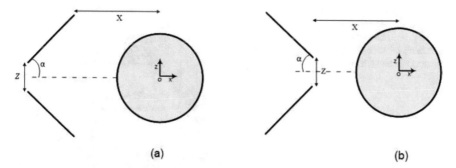

Fig. 4.5 Schematic representation of the (**a**) counterflow inward and (**b**) counterflow outward orientations of the winglet vortex generator pair (Hussain et al. 2016)

transfer and pressure drop were increased with the attack angle of the winglets. They reported that the optimum performance has been obtained for an attack angle of 45°. The orientation of the winglet vortex generator in counterflow inward position was noted to give better results as compared to that of counterflow outward orientation. They concluded that the spanwise heat transfer enhancement was more significant than that in the streamwise direction.

Yoo et al. (1993), Ghorbani-Tari et al. (2013, 2014, 2016), Wang et al. (2012), Henze and Wolfersdorf 2011), Velete et al. (2009), You and Wang (2006), Biswas and Chattopadhyay (1992), Tiggelbeck et al. (1992, 1993), He (2012), Du et al. (2013), Promvonge et al. (2010), Ahmed and Yusoff (2014), Luo et al. (2017), Agarwal and Sharma (2016), Jayavel and Tiwari (2008), Torii et al. (2002), Chu et al. (2009) and Kwak et al. (2005) carried out work on heat transfer enhancement using vortex generators.

4.2   Extended Surface Insert and Tangential Injection Devices

Hilding and Coogan (1964) and Trupp and Lan (1984) provided j and f factors for such extended surface insert. There is a significant contact resistance with this.

Tangential injection devices provide swirl since part of the flow is tangentially injected at locations around the tube circumference at the tube inlet end (Figs. 4.6 and 4.7) (Razgatis and Holman 1976; Dhir et al. 1989; Dhir and Chang 1992).

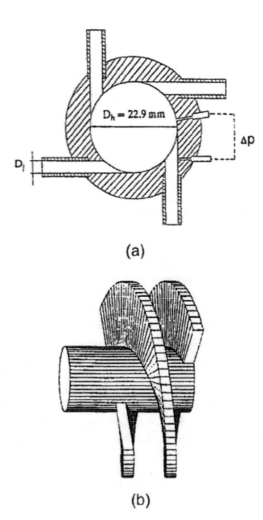

Fig. 4.6 Swirl flow injector designs (**a**) used by Dhir et al. (1989) and (**b**) an alternate injector design (from Dhir and Chang 1992)

(a)

(b)

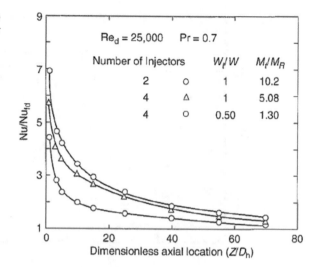

Fig. 4.7 Enhancement ratio for tangential injection of air in a 22.9-mm inside-diameter tube at $Re = 25{,}000$ (from Dhir et al. 1989)

$$\frac{M_t}{M_T} = \frac{W_t^2}{W^2}\frac{A_c}{A_{c,j}} \tag{4.7}$$

$$\frac{Nu}{Nu_{fd}} = 1 + 1.93 \left(\frac{M_t}{M_T}\right)^{0.6} Pr^{-1/7} \exp\left[-m(x/d_i)^{0.6}\right] e \tag{4.8}$$

$$m = 0.89 \left(\frac{M_t}{M_T}\right)^{0.2} Re_d^{-0.18} Pr^{-0.083} \tag{4.9}$$

Sundar and Sharma (2008) carried out an experimental investigation on friction factor and heat transfer characteristics in a circular tube using longitudinal strip inserts. They reviewed the characteristics and performance of full-length longitudinal strip insert inside a circular tube. Rectangular and square cross section strip inserts were used in a circular tube with different aspect ratio (AR). The test section was kept under constant wall heat flux (CWHF), and the flowing fluid had Reynolds number 4000–10,000. Figure 4.8 shows the variation of Nusselt numbers with Reynolds numbers for different aspect ratio and bare tube. It was observed that longitudinal strip with aspect ratio (AR) = 1 gave high heat transfer rate among other different aspect ratios. The following regression equation was made for determining the Nusselt number for longitudinal strip inserts:

$$Nu = 0.02278 Re^{0.824} Pr^{0.412} (AR + 1)^{-0.253} \left(D_i/_l\right)^{0.055} \tag{4.10}$$

Table 4.2 shows the width (l), height (H) and aspect ratio of cross section of strip insert. For checking the accuracy of temperature measurement, the values of Nusselt numbers were calculated experimentally and compared with the predicted values from Sider and Tate equation:

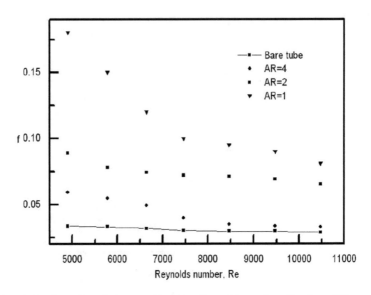

Fig. 4.8 Friction factor for strip inserts with various Reynolds numbers (Sundar and Sharma 2008)

Table 4.2 Dimensions of *l* and *H* (Sundar and Sharma 2008)

Section no.	l (mm)	H (mm)	Aspect ratio (AR) = l/H
1	12	12	1
2	12	6	2
3	12	3	4

$$Nu = 0.036^{0.8} Pr^{0.33} \left(\frac{D}{l}\right)^{0.055} \tag{4.11}$$

Figure 4.9 shows the best fit result for the variation of Nusselt number between experimental result and Sider and Tate equation for water. Figure 4.10 illustrates the variation of friction factor with different aspect ratios. The maximum friction factor of 3.5 was observed for square cross section insert (AR = 1) when compared to the bare tube. Thus, it was cleared from the results that heat transfer rate raised a factor of 22 compared to bare tube at the same Reynolds number ($Re \leq 10{,}000$), and on the other hand, friction factor also raised a factor of about 3.5 at same $Re \leq 10{,}000$. Chen and Hsieh (1992), Hsieh and Huang (2000) and Hsieh and Wen (1996) worked on heat transfer enhancement by using longitudinal strip insert. Another experimental test was carried out on heat transfer performance by using twisted tape in a circular tube by Chen and Hsieh (1992), Hsieh and Huang (2000), Hsieh and Wen (1996), Saha and Dutta (2001), Saha and Langille (2002) and Saha et al. (1989).

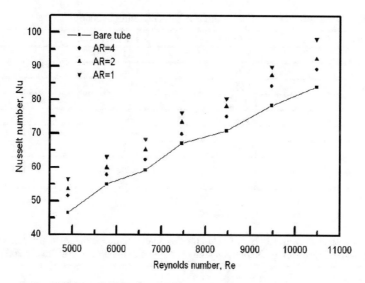

Fig. 4.9 Variation of the Nusselt number for water with and without inserts (Sundar and Sharma 2008)

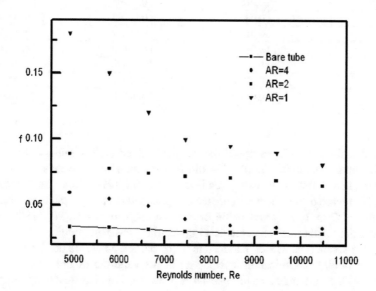

Fig. 4.10 Friction factor for strip insert with various Reynolds numbers (Sundar and Sharma 2008)

Table 4.3 The characteristics of baffle-inserted tubes (Tandiroglu 2006)

Baffle types	β (deg)	H ($\times 10^{-3}$m)	H/D
18,031	180	31	1
18,062	180	62	2
18,093	180	93	3
9031	90	31	1
9062	90	62	2
9093	90	93	3
4531	45	31	1
4562	45	62	2
4593	45	93	3

Table 4.4 Empirical correlation constant of the time-averaged irreversibility distribution ratio as a function of the irreversibility distribution ratio at steady state conditions (Tandiroglu 2006)

	$\phi_t = \phi_\infty + Be^{\frac{-Re}{C}}$		
Baffle types	ϕ_∞	B	C
18,031	0,00224	0,00804	5027,96314
18,062	0,00471	0,05370	3796,02346
18,093	0,01602	0,06305	4791,80677
9031	0,00627	0,01445	6058,14220
9062	0,01622	0,05627	5014,31544
9093	0,02258	0,08473	4828,38769
4531	0,00386	0,01386	5132,66612
4562	0,01253	0,02376	7044,41851
4593	0,01376	0,07642	4878,08305

Tandiroglu (2006) investigated the transient forced convection heat transfer of circular tube with baffle inserts. He did experimental work to enhance the heat transfer characteristics by using semi-circular baffle plates insert in a turbulent flow. He reported the effect of geometrical parameters such as pitch to tube inlet diameter $H/D = 1, 2, 3$ and baffle orientation angle $\beta = 45°, 90°, 180°$ on the performance of thermal characteristics under the transition conditions. Table 4.3 shows the detailed geometric parameters of the baffled tubes. He used the principle of irreversibility minimization and compared the variation of time-averaged irreversibility distribution ratio with nine baffle plate inserts and presented it in the form of graphs.

Empirical correlation of time-averaged irreversibility distribution ratio as a function irreversibility distribution ratio had been tabulated in the Table 4.4 at steady state conditions for all nine baffle plate inserts. Table 4.5 shows the empirical correlation between time-averaged Nusselt number and the Nusselt number for steady state condition for all nine baffle inserts. Table 4.6 shows the empirical

Table 4.5 Empirical correlation constant of the time-averaged Nusselt number as a function of the steady state Nusselt number (Tandiroglu 2006)

| Baffle types | $Nu_t = Nu_\infty + Be^{\frac{-Re}{C}}$ | | |
	Nu_∞	B	C
SMOOTH	13,84546	3,368340	6223,43534
18,031	52,87346	14,76116	5949,25628
18,062	115,1245	10,45953	7620,69965
18,093	71,23456	17,81594	6169,92327
9031	122,7171	59,20562	4690,94526
9062	96,71809	19,39754	6574,59198
9093	70,68195	17,07486	6236,93904
4531	81,70303	28,78974	5442,94480
4562	65,16575	30,39422	4774,67246
4593	66,73515	8,373950	7265,75080

Table 4.6 Empirical correlation constant of the time-averaged friction factor as a function of the steady state friction (Tandiroglu 2006)

| Baffle types | $f_t = f_\infty + Be^{\frac{-Re}{C}}$ | | |
	f_∞	B	C
SMOOTH	0,00053	0,00149	4590,53527
18,031	0,18350	0,11285	5428,44683
18,062	0,08435	0,10367	5829,43319
18,093	0,02722	0,01509	5399,10468
9031	0,07818	0,01502	5246,13616
9062	0,02656	0,00844	5295,39708
9093	0,01916	0,00878	5354,86585
4531	0,12252	0,08009	5447,87732
4562	0,03703	0,00473	5222,26911
4593	0,02165	0,00762	5309,48407

correlation between time-averaged friction factor as a function of steady state friction factor for all nine baffle inserts. Figure 4.11 shows the variation in Nusselt number for steady state and unsteady state condition with respect to the Reynolds number for smooth tube and all nine baffle inserted tubes. Figure 4.12 shows the variation friction factor for steady state and unsteady state flow conditions with respect to the Reynolds number for smooth tube and all nine baffle inserted tubes.

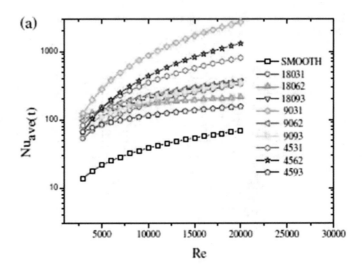

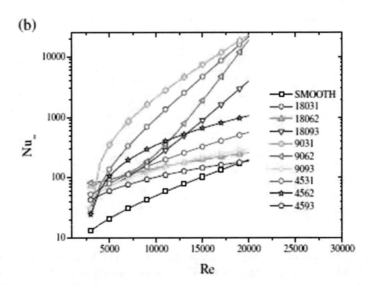

Fig. 4.11 (a) The time-averaged Nusselt number with respect to the Reynolds number. (b) Nusselt number vs. Reynolds number for steady state flow conditions (Tandiroglu 2006)

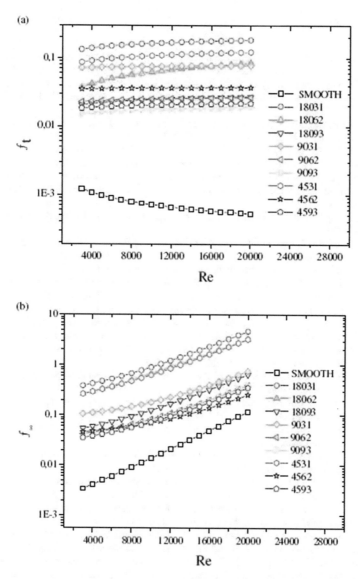

Fig. 4.12 (a) The time-dependent friction factor f with respect to the Reynolds number, (b) friction factor with respect to the Reynolds number for steady state flow (Tandiroglu 2006)

References

Agarwal S, Sharma RP (2016) Numerical investigation of heat transfer enhancement using hybrid vortex generator arrays in fin-and-tube heat exchangers. ASME J Therm Sci Eng Appl 8:031007

Ahmed HE, Yusoff MZ (2014) Impact of delta-winglet pair of vortex generators on the thermal and hydraulic performance of a triangular channel using Al_2O_3–water nano-fluid. ASME J Heat Transf 136:021901

Biswas G, Chattopadhyay H (1992) Heat transfer in a channel with built-in wing-type vortex generators. Int J Heat Mass Transf 35(4):803–814

Chen JD, Hsieh SS (1992) Buoyancy effect on the laminar forced convection in a horizontal tube with a longitudinal thin plate insert. Int J Heat Mass Transf 35(1):263–267

Chu P, He YL, Tao WQ (2009) Three-dimensional numerical study of flow and heat transfer enhancement using vortex generators in fin-and-tube heat exchangers. ASME J Heat Transf 131:091903

Dhir VK, Chang F (1992) Heat transfer enhancement using tangential injection. ASHRAE Trans 98(2):383–390

Dhir VK, Tune VX, Chang F, Yu J (1989) Enhancement of forced convection heat transfer using single and multi-stage tangential injection. In: Goldstein RJ, Chow LC, Anderson EE (eds) Heat transfer in high energy heat flux applications, ASME Symp. HTD, vol 119, pp 61–68

Du X, Feng L, Yang Y, Yang L (2013) Experimental study on heat transfer enhancement of wavy finned flat tube with longitudinal vortex generator. Appl Therm Eng 50:55–62

Eiamsa-ard S, Thianpong C, Eiamsa-ard P, Promvonge P (2009) Convective heat transfer in a circular tube with short-length twisted tape insert. Int Commun Heat Mass Transf 36:365–371

Ghorbani-Tari Z, Wang L, Sunden B (2013) Endwall convective heat transfer around a single bluff body in rectangular channel. In Proc. of 8th world conf. on experimental heat trans., fluid mechanics and thermodynamics, Lisbon, Portugal 316

Ghorbani-Tari Z, Wang L, Sunden B (2014) Heat transfer control around an obstacle by using ribs in the downstream region. ASME J Therm Sci Eng 6:041010

Ghorbani-Tari Z, Wang L, Sunden B (2016) Heat transfer characteristics around an obstacle controlled by the presence of ribs. Heat Transfer Res 47:893–906

He J (2012) Vortex-enhanced heat transfer by a new delta-winglet array. Ph.D. Thesis, University of Illinois, Urbana-Champaign

Henze M, Wolfersdorf JV (2011) Influence of approach flow conditions on heat transfer behind vortex generators. Int J Heat Mass Transf 54:279–287

Hilding WE, Coogan CH Jr (1964) Heat transfer and pressure drop in internally finned tubes. In: ASME symposium on air cooled heat exchangers. ASME, New York, pp 57–84

Hsieh SS, Huang IW (2000) Experimental studies for heat transfer and pressure drop of laminar flow in horizontal tubes with/without longitudinal inserts

Hsieh SS, Wen, M-Y (1996) "Developing three-Dimensional laminar mixed convection in a circular tube inserted with longitudinal Strips," Int J Heat Mass Transf 39:299–310

Hussain S, Liu J, Wang L, Sunden B (2016) Effects on endwall heat transfer by a winglet vortex generator pair mounted upstream of a cylinder. J Enhanc Heat Transf 23(3):241–262

Jayavel S, Tiwari S (2008) Numerical study of flow and heat transfer for flow past inline circular tubes built in a rectangular channel in the presence of vortex generators. Numer Heat Transf Part A 54:777–797

Kwak KM, Torii K, Nishino K (2005) Simultaneous heat transfer enhancement and pressure loss reduction for finned-tube bundles with the first or two transverse rows of built-in winglets. Exp Thermal Fluid Sci 29:625–632

Luo L, Wen F, Wang L, Sunden B, Wang S (2017) On the solar receiver thermal enhancement by using dimple combined with delta winglet vortex generator. Appl Therm Eng 111:586–598

Martemianov S, Okulov VL (2004) On heat transfer enhancement in swirl pipe flows. Int J Heat Mass Transf 47(10–11):2379–2393

Moawed M (2011) Heat transfer and friction factor inside elliptic tubes fitted with helical screw-tape inserts. J Renew Sustain Energy 3:1–15

Promvonge P, Chompookham T, Kwankaomeng S, Thianpong C (2010) Enhanced heat transfer in a triangular ribbed channel with longitudinal vortex generator. Energy Conserv Manage 51 (6):1242–1249

Razgatis R, Holman JP (1976) A survey of heat transfer in confined swirl flows. Heat Mass Transf Process 2:831–866

Saha SK, Dutta A (2001) Thermohydraulic study of laminar swirl flow through a circular tube fitted with twisted tapes. J Heat Transf 123:417–427

Saha SK, Langille P (2002) Heat transfer and pressure drop characteristics of laminar flow through a circular tube with longitudinal strip inserts under uniform wall heat flux. J Heat Transf 124 (3):421–432

Saha SK, Gaitonde UN, Date AW (1989) Heat transfer and pressure drop characteristics of laminar flow in a circular U tube fitted with regularly spaced twisted-tape elements. Exp Thermal Fluid Sci 2(3):310–322

Sarac BA, Bali T (2007) An experimental study on heat transfer and pressure drop characteristics of decaying swirl flow through a circular pipe with a vortex generator. Exp Thermal Fluid Sci 32:158–165

Sundar LS, Sharma KV (2008) Experimental investigation of heat transfer and friction factor characteristics in a circular tube with longitudinal strip inserts. J Enhanc Heat Transf 15 (4):325–333

Tandiroglu A (2006) Irreversibility minimization analysis of transient heat transfer for turbulent flow in a circular tube with baffle inserts. J Enhanc Heat Transf 13(3):215–229

Tiggelbeck S, Mitra N, Fiebig M (1992) Flow structure and heat transfer in a channel with multiple longitudinal vortex generators. Exp Thermal Fluid Sci 5:425–436

Tiggelbeck S, Mitra NK, Fiebig M (1993) Experimental investigations of heat transfer enhancement and flow losses in a channel with double rows of longitudinal vortex generators. Int J Heat Mass Transf 36:2327–2337

Torii K, Kwak KM, Nishino K (2002) Heat transfer enhancement accompanying pressure-loss reduction with winglet-type vortex generators for fin-tube heat exchangers. Int J Heat Mass Tranf 45:3795–3801

Trupp AC, Lau ACY (1984) Fully developed laminar heat transfer in circular sector ducts with isothermal walls. J Heat Transf 106:467–469

Velete CM, Hansen MOL, Okulov VL (2009) Helical structure of longitudinal vortices embedded in turbulent wall-bounded flow. J Fluid Mech 619:167–177

Wang L, Salewski M, Sunden B, Borg A, Abrahamsson H (2012) Endwall convective heat transfer for bluff bodies. Int Commun Heat Mass Transf 39:167–173

Yoo SY, Goldstein RJ, Chung MK (1993) Effects of angle of attack on mass transfer from a square cylinder and its base plate. Int J Heat Mass Transf 36:371–381

You D, Wang M (2006) Large-eddy simulations of longitudinal vortices embedded in a turbulent boundary layer. AIAA J 44:3032–3039

Chapter 5
Numerical Simulation of Integral Roughness, Laminar Flow in Tubes with Roughness and Reynolds Analogy for Heat and Momentum Transfer

5.1 Numerical Simulation of Integral Roughness

Both 2D and 3D roughness involve complications, and solving momentum and energy equations for complex flow geometries is a different task because of local flow separations, etc. However, several bold attempts have been made by researchers Benodekar et al. (1985), Hung et al. (1987), Fodemski and Collins (1988), Durst et al. (1988), Becker and Rivir (1989), Fujita et al. (1986), Hijikata et al. (1987), Prakash and Zerkle (1995), Ekkad and Han (1997), Iacovides and Raisee (1999), Rigby et al. (1997) and Patankar (1990).

Chen and Patel (1988) gave a two-layer turbulence model and divided the flow into near-wall region and fully developed core region. The Chen and Patel (1988) model used standard k-ε model in the core region, and one equation model was used by Arman and Rabas (1991, 1992) and Rabas and Arman (1992) for the prediction of 2D roughness geometries (Fig. 5.1). This 2D rib inserted turbulent flow near-wall consisted of stable recirculation and boundary layer development zone. Above this zone, there was a free shear region full of large eddies. However, the k-ε two-equation model failed to do a good job to find the reattachment length. The friction velocity changed sign at the reattachment point, and the wall shear stress varied with axial position.

Figure 5.2 shows the predicted friction factors and wall shear stress composed of data of Webb et al. (1971). Figure 5.3 shows a comparison between the predicted and experimental heat transfer data of Webb et al. (1971). Iacovides and Raisee (1999) explored turbulence modelling issues related to the flow and heat transfer in a square channel with rib-roughened walls. They compared four turbulence models: a zonal k-ε, a low-Re k-ε, a zonal differential stress model (DSM) and a low-Re DSM. They compared the models differently with the experimental data. Arman and Rabas (1992) studied numerically the effect of rib shape on Nusselt number and friction

© The Author(s), under exclusive license to Springer Nature Switzerland AG 2020
S. K. Saha et al., *Insert Devices and Integral Roughness in Heat Transfer Enhancement*, SpringerBriefs in Applied Sciences and Technology,
https://doi.org/10.1007/978-3-030-20776-2_5

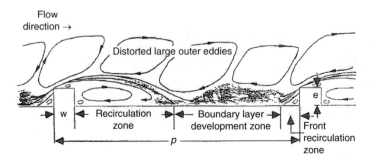

Fig. 5.1 Structure of turbulent flow over a two-dimensional rib (from Annan and Rabas 1991)

factor and compared their predicted data with that experimentally obtained by Hijikata and Mori (1987) and Nunner (1956) (Table 5.1 and Fig. 5.4).

Chaube et al. (2006) investigated the effect of roughened surface on the heat transfer and flow friction characteristics of solar air heater. A two-dimensional analysis was done, and a correlation for maximum heat transfer with minimum pressure penalty has been established. They computationally investigated ten different types of rib shapes in the range of Reynolds number 3000–20,000. Simulation was carried out for transitional flow regime ($5 \leq e^+ \leq 70$) and fully rough regime ($e^+ \geq 70$). They selected a turbulence model for comparing different turbulence model available in literature.

Figure 5.5 shows the variation of heat transfer enhancement with Reynolds number for ten different geometries of rough surfaces at constant pumping power in transitionally rough flow regime. Figure 5.6 illustrates the comparison of heat transfer index of ten different geometries of rough surfaces at constant pumping power in fully rough flow regime. It was observed that in fully rough flow, heat transfer increment over that of transitional flow regime was only marginal but friction factor increased more than two times.

The semi-empirical discrete-element method was used by Taylor et al. (1984, 1988) and Taylor and Hodge (1992) for 3D roughness. Assuming linear problem and parabolic boundary layer equation, the solution to the momentum equations provided the velocity profiles and the friction factor, and thereafter, the energy equation was solved using previously obtained velocity profiles. The drag coefficient and the heat transfer from the discrete roughness elements were accounted for by the empirical expressions. In these numerical methods, the Prandtl mixing length model with Van Driest damping and constant turbulent Prandtl number has been used, wherever necessary. The pressure gradient effects on the turbulence were accounted for using a wall parameter, as suggested by Kays and Crawford (1980).

Friction factor correlations of Zukauskas (1972) for tube banks were used by reinterpreting it as a drag coefficient. Taylor and Hodge (1992) used discrete-element model to compare the 3D roughness data of Dipprey and Sabersky (1963), Takahashi et al. (1988), Gowen and Smith (1968) and Cope (1945). Taylor and Hodge (1992) studied numerically the effect of different spherical shapes and

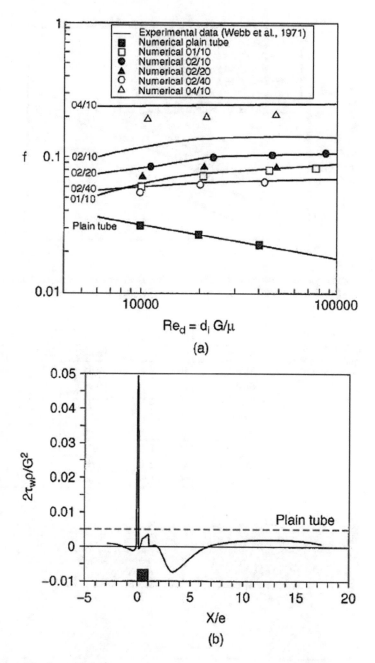

Fig. 5.2 (**a**) Predicted friction factors compared with data of Webb et al. (1971); (**b**) predicted axial variation of wall shear stress for $e/d_i = 0.01$ and $p/e = 20$ at $Re = 47,000$ (from Arman and Rabas 1991)

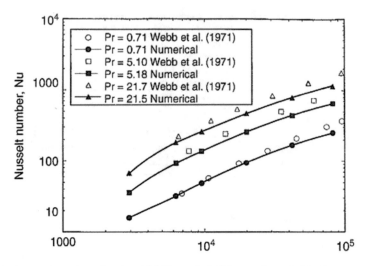

Fig. 5.3 Comparison of predicted and Webb et al. (1971) experimental data for transverse rib roughness ($e/d_i = 0.02$, $p/e = 20$) for $0.71 \leq Pr \leq 21.7$ (Rabas and Arman 1992)

Table 5.1 Comparison of mean Nusselt numbers and efficiency index ($e/d_i = 0.02$) (Webb and Kim 2005)

Shape	p/e	$Re = 9400$		$Re = 39,000$	
		Nu/Nu_p	η/η_{trap}	Nu/Nu_p	η/η_{trap}
Sine	10	2.17	0.95	2.22	0.96
Sine	20	1.84	0.97	1.88	0.99
Semicircle	10	2.13	0.93	2.21	0.96
Semicircle	20	1.79	0.95	1.83	0.96
Arc	10	1.98	0.86	2.16	0.91
Arc	20	1.71	0.91	1.78	0.94
Trapezoid	10	2.29	1.00	2.32	1.00
Trapezoid	20	1.89	1.00	1.91	1.00

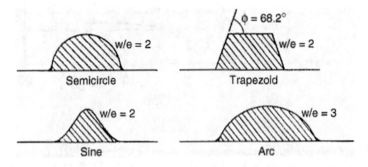

Fig. 5.4 Different rib shape analysis (Arman and Rabas 1992)

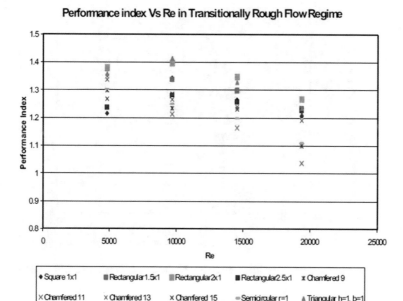

Fig. 5.5 Variation of heat transfer enhancement for constant pumping power requirement with Reynolds number (Chaube et al. 2006)

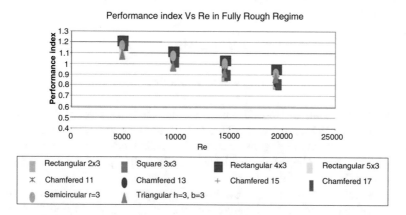

Fig. 5.6 Variation of heat transfer enhancement at constant pumping power with Reynolds number (Chaube et al. 2006)

spacing, and significant performance improvements were possible with an increased roughness curvature and closer roughness spacing. They worked with boundary layer flow on rough plates. Chakroun and Taylor (1992) treated accelerating boundary layer flow over rough plates. James et al. (1994) extended the discrete-element model to 2D rib roughness geometry. They have deduced a so-called blockage factor

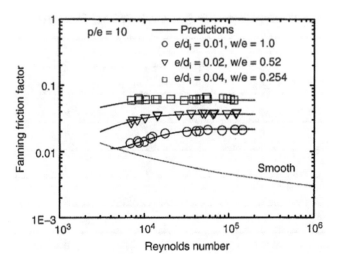

Fig. 5.7 Comparison of predicted friction factors with the data of Webb et al. (1971) (James et al. 1994)

based on the experimental measurements by Faramarzi and Logan (1991) and Mantle (1966) on a ribbed geometry.

Equations give the blockage factor and the friction factor and Nusselt number correlations.

$$\beta_x = \beta_y = 1 - {}^{[w+(\gamma+1)(e-y)]}\!\big/_p \tag{5.1}$$

$$Nu_{rib} = 0.023 E_r Re_d^{0.8} Pr^{0.4} \tag{5.2}$$

James et al. (1994) also modelled the prediction method for repeated rib geometry of Webb et al. (1971), Berger and Han (1979), Mendes and Mauricio (1987) and Baughn and Roby (1992). Figure 5.7 shows the prediction for $p/e = 10$ tube of Webb et al. (1971). Extension of the model to spirally ribbed geometry is possible by considering the effect of flow swirl.

Li et al. (2009) studied the effect of internally roughened dimpled tube on the heat transfer enhancement in the Reynolds number range of 2000–11,000. They have done three-dimensional simulation on five samples to study the heat transfer enhancement. They determined the fluid flow and heat transfer characteristics by calculating the local and overall friction factors, Nusselt number and Colburn j factor. The synergy principle described the effect of dimples on the heat transfer performance. Arrangement of the dimples had very little effect (within 2%) on the heat transfer but performance of heat transfer was significantly affected by the size of the dimples. Therefore, they investigated the tube with optimum dimple size. Geometrical parameters of five dimpled tubes and smooth tube were summarized in Table 5.2.

Table 5.2 Dimensional configuration of the tubes (Li et al. 2009)

Physical quantity	Smooth tube	Dimpled tubes				
	Tube 0	Tube 1	Tube 2	Tube 2*	Tube 4	Tube 5
D, mm	16	16	16	16	16	16
L, mm	320	320	320	320	320	320
A, mm^2	7787	7796	7961	7961	8459	8658
Arrangement		In-lined	In-lined	Staggered	In-lined	In-lined
p, mm		3.6	2	2	1	0.42
h, mm		0.1	0.5	0.5	1	1.5
d, mm		1.4	3	3	4	4.58
r, mm		2.5	2.5	2.5	2.5	2.5

Table 5.3 Boundary condition for numerical simulations (Li et al. 2009)

Boundary	Quantity		USED	Dimpled region		DSED
				Fluid	Tube	
Inlet	Velocity	u	$u = u_{in}$			
		v	$v = 0$			
		w	$w = 0$			
	Temperature T		$T = T_{in}$			
Inlet	Velocity	u	$\frac{\partial u}{\partial y} = 0$	$\frac{\partial u}{\partial y} = 0$		$\frac{\partial u}{\partial y} = 0$
		v	$v = 0$	$v = 0$	$v = 0$	$v = 0$
		w	$\frac{\partial w}{\partial y} = 0$	$\frac{\partial w}{\partial y} = 0$	$w = 0$	$\frac{\partial w}{\partial y} = 0$
	Temperature T		$\frac{\partial T}{\partial y} = 0$	$\frac{\partial T}{\partial y} = 0$	$T = T_w$	$\frac{\partial T}{\partial y} = 0$
Outlet	Velocity	u				$\frac{\partial u}{\partial x} = 0$
		v				$\frac{\partial v}{\partial x} = 0$
		w				$\frac{\partial w}{\partial x} = 0$
	Temperature T					$\frac{\partial T}{\partial x} = 0$

Table 5.3 shows the boundary conditions required for numerical simulation for the three regions (upstream extended region, dimpled region, downstream extended region). USED means the upstream extended domain, and DSED means the downstream extended domain. Figures 5.8 and 5.9 describe the variation of normalized Nusselt number and normalized friction factor, respectively, with Reynolds number and compared the simulated results with the experimental results, respectively. Figure 5.10 shows the effect of dimple sizes on the normalized Nusselt number. The heat transfer rate decreased with the increase in Reynolds number with bigger dimples. Figure 5.11 shows the variation of normalized fiction factor with Reynolds number. The normalized friction factor increases with an increase in Re for the tubes with smaller dimples.

Song et al. (2013) carried out numerical investigation to study the forced convection heat transfer enhancement characteristics of flow in a ribbed channel which

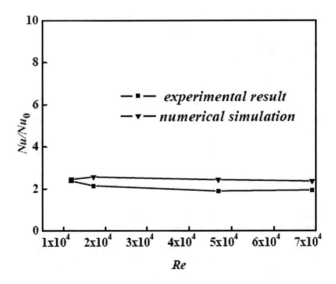

Fig. 5.8 Computed and measured normalized Nusselt number at different Reynolds number for the smooth and dimpled cases (Li 2009)

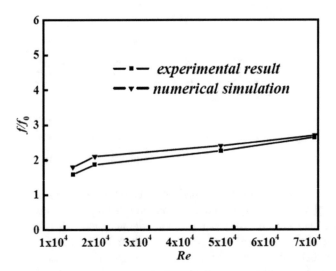

Fig. 5.9 Computed and measured normalized friction factor at different Reynolds number for the smooth and dimpled cases (Li 2009)

has different shaped deflectors installed. Different mechanisms of heat transfer enhancement by disturbing the boundary layer using ribbed configurations have been presented by Han et al. (1978), Ligrani et al. (2003), Ligrani (2013), Bergles and Manglik (2013), Han and Park (1988), Han and Zhang (1991), Han et al. (1985,

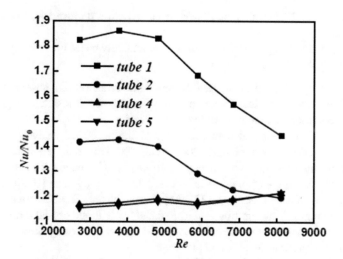

Fig. 5.10 Effect of different dimple sizes (tubes 1, 2, 4 and 5) on the normalized Nusselt number vs. *Re* (Li 2009)

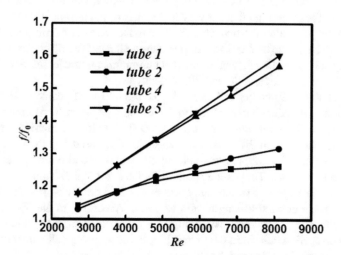

Fig. 5.11 Effect of different dimple sizes (tubes 1, 2, 4 and 5) on the normalized friction factor vs. *Re* (Li 2009)

1989) and Han (1988). The general observations on ribs from the literature are as follows:

(a) Performance of angled ribs is better compared to that of transverse ribs.
(b) Mass transfer for parallel ribs is greater than that for crossed ribs.
(c) 45° attack angle shows effective increase in heat transfer.

(d) V-shaped and delta-shaped ribs showed superior performance over that of angled ribs.
(e) Detached and staggered ribs perform comparatively better than symmetrical and attached ribs (Kanoun et al. 2011).

Rau et al. (1998), Jia et al. (2002) and Kiml et al. (2003) observed that the disturbance of side boundary layer by the presence of rib resulted in generating streamwise vorticity as the flow rolled up by itself.

Li et al. (2013) presented both experimental and numerical analysis on thermo-hydraulic characteristics of channel flow having ribs with rectangular cross section. Continuous transverse ribs with large pitch-to-height ratios have been considered for the analysis. The range of pitch-to-height ratio has been given as 10–30. Turbulent flow regime with $57,000 < Re < 127,000$ was considered. The heat transfer coefficients in the region between ribs were calculated using liquid crystal thermography. For numerical analysis, the Reynolds-averaged Navier–Stokes equations have been employed to represent the physical problem.

The realizable k-ε turbulence model has been used. Their study mainly focussed on the thermo-hydraulic performance in the region between the first repeated ribs as the flow field, and thermal field in this particular region was not periodically fully developed. They observed that for very low values of pitch-to-height ratio, the flow reattachment point after the first rib cannot be found. Also, if the rib pitch-to-height ratio is very large, the heat transfer augmentation effect of the ribs becomes ineffective. They concluded that the maximum and minimum pressure losses were observed for pitch-to-height ratios of 20 and 30, respectively.

Ajeel et al. (2019) numerically investigated the impact of geometrical parameter design on heat transfer characteristics to achieve compactness. The numerical study was targeted in Reynolds number range 10,000–30,000 for symmetrical trapezoidal corrugated channel with SiO_2-water as nanofluid. They used finite volume method, the SIMPLE algorithm for pressure–velocity fields and second-order upwind scheme for convective terms. They numerically simulated for 0–8% volume fraction of nanofluids. The objective of the study was to evaluate the effect of height-to-width ratio variation and pitch-to-length ratio variation. According to the obtained simulated results, they concluded that fixing one parameter height-to-width ratio and varying corrugation ratio from 0.075 to 0.175, the average Nusselt number increased with increase in Reynolds number for every considered case.

Also, decreased P/L ratio has resulted in increased average Nusselt number as Reynolds number increased, and it was found that 15.29% increment resulted at Reynolds number 30,000. These results can be understood by the fact that strong circulation zones were near corrugations. They examined the impact of corrugation ratio on pressure drop and observed that it was higher for higher pitch ratio. Results revealed the effect of nanofluids on the Nusselt number and observed that Nusselt number enhancement decreased with increased Reynolds number. Also, PEC diminished with increased value of Reynolds number, and it was 9.22%, 8.788% and 8.71% decrement for $p/l = 0.075$, 0.125 and 0.175, respectively, for the considered Reynolds number. The highest PEC value was 2.9 at lowest ratio of p/l calculated for Reynolds number 10,000.

They evaluated height-to-width ratio (h/W) impact on heat transfer characteristics and observed that average Nusselt number increased as h/W increased. They numerically estimated 99.45% increment as h/W ratio varies from 0.0 to 0.05 at fixed Reynolds number 30,000. For $h/W = 0$, mean flat plate offers lowest pressure drop. The highest enhancement of Nusselt number was achieved at $h/W = 0.05$ for Reynolds number 10,000. The highest PEC value was obtained at $h/W = 0.05$. PEC decreased with increase in Reynolds number, and it was observed that 1.82%, 9.11%, 8.66% and 9.22% reduction at $h/W = 0$, 0.03, 0.04 and 0.05, respectively, in considered range of Reynolds number. Ajeel et al. (2019) proposed that height-to-width ratio (h/W) of 0.05 conjugated with P/L of 0.075 was the optimum design parameter.

5.2   Laminar Flow in Tubes with Roughness

Several factors influence laminar flow in tubes. These are thermal boundary conditions, entrance region effects and natural convection. Vicente et al. (2002a) obtained laminar flow data in dimpled tube geometry (Figs. 5.12, 5.13, 5.14 and Table 5.4). The correlations developed by Vicente et al. (2002a) are

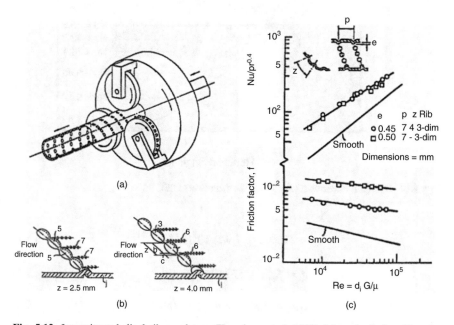

Fig. 5.12 Intermittent helical rib roughness (Kuwahara et al. 1989) (**a**) method of making the interior dimples, (**b**) effect of transverse dimple pitch on local vortices and (**c**) performance of intermittent helical ribs. Tubes with similar dimple configurations were tested by Vicente et al. (2002a)

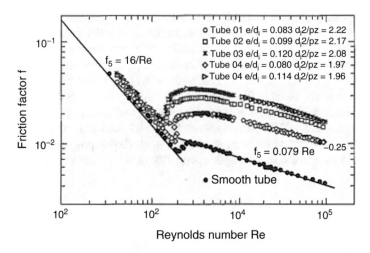

Fig. 5.13 Friction factor vs. Reynolds number for five dimpled tubes (Vicente et al. 2002a)

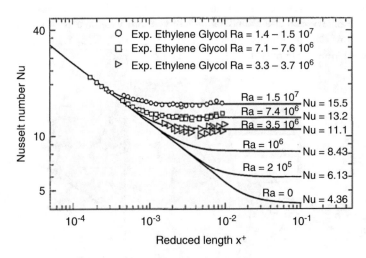

Fig. 5.14 Nusselt number vs. x'' for Tube 02 (Vicente et al. 2002a)

$$Re_{\text{crit}} = 2100\left[1 + 7.9 \times 10^7 (e/d_i)^{-6.54}\right]^{-0.1} \tag{5.1}$$

$$f = f_s\left[1 + 123.2\left(^e/_{d_i}\right)^{2.2} Re_d^{0.2}\right] \tag{5.2}$$

$$Nu_d = 4.36[1 + Ra/67,000]^{0.24} \tag{5.3}$$

Esen et al. (1994b) and Obot et al. correlated data in the form given by

Table 5.4 Roughness dimension of dimpled tubes tested (Vicente et al. 2002a)

Tube no.	d_i (mm)	e (mm)	p (mm)	z (mm)	e/d_i	d_i^2/pz
01	16.0	1.33	13.0	8.85	0.0831	2.225
02	16.0	1.58	13.1	8.99	0.0988	2.175
03	16.0	1.91	13.8	8.89	0.1194	2.085
04	16.0	1.28	14.6	8.91	0.0800	1.975
05	16.0	1.83	14.5	9.02	0.1144	1.964
06	16.0	1.59	17.2	9.02	0.0944	1.652
07	16.0	1.84	16.6	9.06	0.1150	1.700
08	16.0	1.87	16.8	8.90	0.1169	1.709
09	16.0	1.55	10.9	8.90	0.0969	2.646
10	16.0	1.64	11.4	8.76	0.1025	2.575

$$Re_m = \left(^{Re_{c,s}}/_{Re_c}\right)Re \tag{5.4}$$

$$f_m = \left(^{f_{c,s}}/_{f_c}\right)f \tag{5.5}$$

$$Nu_m = \left(^{Nu_{c,s}}/_{Nu_c}\right)Nu \tag{5.6}$$

They have covered a wide Prandtl number range. Normalized data are shown in Fig. 5.15. Similarity parameters have been utilized in their work. The laminar flow heat transfer data are correlated as

$$Nu_d = 0.008fRe_d^{1.5}Pr^{0.4} \tag{5.7}$$

They also attempted to provide similar correlation in the turbulent region.

Integral roughness has been used in low-flow automotive radiators (Farrell et al. 1991; Olsson and Sunden 1996) (Fig. 5.16, Table 5.5). The data provide the average heat transfer coefficient over the channel length. Figure 5.17 shows the f and j characteristics versus Reynolds number. Smooth tube, full ribbed tube, broken ribbed tube, dimple tube and OSF inserts have been investigated. Transition region data have been collected. Olsson and Sunden (1998a, b) investigated rib-roughened rectangular channels (with fractional aspect ratio) having cross-ribs, parallel ribs, cross V ribs, parallel V ribs and multiple V ribs. Figure 5.18 shows a sketch of the channel with multiple V ribs. Esen et al. (1994a) and Vicente et al. (2002b) have studied the friction factor slope of the enhanced tubes.

Xie et al. (2013) numerically studied heat transfer augmentation in ribbed channels with square cross-sectioned ribs having differently positioned deflectors. Kanoun et al. (2011) Al-Qahtani et al. (2002), Saha and Acharya (2005), Viswanathan and Tafti (2006) carried out similar numerical studies.

The effect of rib installations in reciprocating channels for heat transfer augmentation was presented by Perng and Wu (2013). They carried out a numerical analysis and studied the effect of rib–pitch, rib–land and rib–blockage ratios on heat transfer

Fig. 5.15 The normalized
friction factor (**a**) or
normalized Nusselt number
(**b**) vs. normalized Reynolds
number for commercial
enhanced tubes as reported
by Esen et al. (1994b) and
Obot et al. (2001)

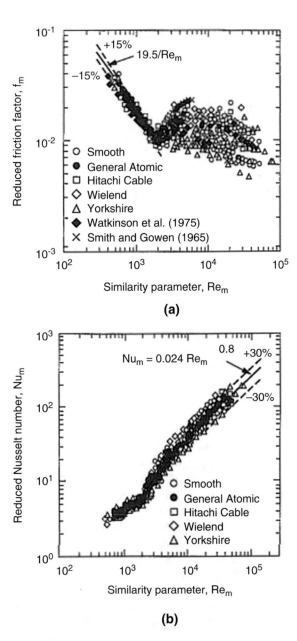

Fig. 5.15 The normalized friction factor (**a**) or normalized Nusselt number (**b**) vs. normalized Reynolds number for commercial enhanced tubes as reported by Esen et al. (1994b) and Obot et al. (2001)

in turbulent flow. They observed 94.44–184.65% increase in average time-mean Nusselt number in the ribbed channel over that in a smooth/plain channel which is subjected to reciprocating motion. They attributed this increase in heat transfer to the combined effect of buoyancy momentum of fluid at the inlet and the reciprocating forces. They concluded that the heat transfer from the wall to the turbulent flow was

Fig. 5.16 Outside surface
geometry of the radiator
tubes tested by Olsson and
Sunden (1996). From the
top d5, d4, d3, d2, r2, rl and
s (Olsson and Sunden 1996)

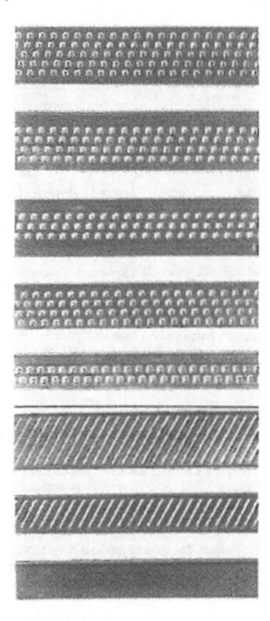

observed to increase with the increase in rib–land and rib–blockage ratios, and the
reverse trend was observed for the increase in rib–pitch ratio.

Lee et al. (2012) studied the effect of dimple arrangement on heat transfer
enhancement in a dimpled channel turbulent flow. They concluded that the staggered
array of dimples showed better enhancement performance than that of in-line array
of dimples. In the dimpled channels, the flow separation and reattachment takes
place inside the dimple. Thus, heat transfer augmentation can be observed around the

Table 5.5 Geometric dimensions of the radiator tubes tested (Olsson and Sunden 1996)

Tube	D_h (mm)	p (mm)	e (mm)	H (mm)	θ (degree)	W (mm)
Smooth (s)	3.11	–	–	1.74	–	14.5
Rib-roughened (r1)	3.11	4.0	0.18	1.74	30	14.6
Rib-roughened (r2)	3.05	4.0	0.20	1.64	30	21.6
Dimpled (d1)	3.53	4.5	0.45	2.00	30	15.2
Dimpled (d2)	2.81	4.8	0.45	1.52	18	18.3
Dimpled (d3)	3.08	4.8	0.45	1.65	18	23.2
Dimpled (d4)	3.08	4.8	0.45	1.65	18	23.2
Dimpled (d5)	3.08	4.8	0.45	1.65	18	23.2
OSF	D_h (mm)	s (mm)	b (mm)	L_p (mm)	H (mm)	W (mm)
(osf1)	2.82	3.5	2.6	5.0	3.0	26.2
(osf2)	2.82	3.5	2.6	5.0	3.1	43.4

reattachment point. The portion of the reattached flow going out of the dimple increases the fluid mixing by interacting with the flow passing over the dimple. The vortex pairs which are generated at the downstream of the dimple flow along the direction of the dimple diagonal are responsible for the enhancement of the heat transfer. Based on large eddy simulation, Lee et al. (2008) explained that the vortices which are generated at the shear layer, as a consequence of separation of the flow play a vital role in heat transfer augmentation.

5.3 Reynolds Analogy for Heat and Momentum Transfer

Correlations for turbulent flow friction and heat transfer characteristics of rough surfaces have been developed for geometrically similar roughness. Correlations of friction factor and Nusselt number have been developed, and these can be applied to any arbitrary family of geometrically similar roughness. Friction similarity law of Nikuradse (1933) and the works of Schlichting (1979) have been used to develop the correlations. The velocity profile is based on the law of the wall velocity distribution for rough surfaces (Hinze 1975; Schlichting 1979). Figure 5.19 may be interpolated for critical phenomenological data analysis.

$$\frac{u}{u^*} = 2.5 \ln \frac{y}{e} + B(e^+) \tag{5.8}$$

$$e^+ = \frac{eu^*}{v} = \frac{e}{d_i} Re_d \sqrt{\frac{f}{2}} \tag{5.9}$$

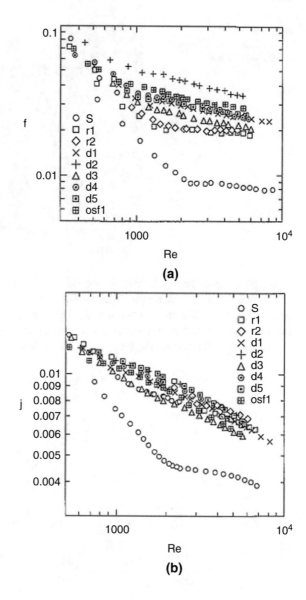

Fig. 5.17 Friction factor (**a**) and *j* factor (**b**) vs. Reynolds number for the radiator tubes tested by Olsson and Sunden (1996)

$$\tau_0 = -\frac{d_i dp}{4\rho\, dx} \tag{5.10}$$

$$\frac{\bar{u}}{u^*} = \left(\frac{2}{f}\right)^{1/2} = -2.5\ln\left(\frac{2e}{d_i}\right) - 3.75 + B(e^+) \tag{5.11}$$

These equations hold for any family of geometrically similar roughness. However, the function $B(e^+)$ will be different for different basic roughness types.

Fig. 5.18 Multiple
V-ribbed channel (Olsson
and Sunden 1998b)

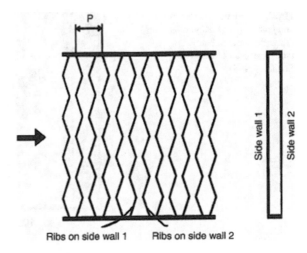

Ribs on side wall 1   Ribs on side wall 2

Dipprey and Sabersky (1963) developed correlation based on the heat–momentum transfer analogy for rough surfaces. This model can be applied to any type of geometrically similar surface roughness. The function for correlation is obtained from momentum and energy equations for turbulent flow and integrating across the boundary layer thickness. The basic model is explained by Fig. 5.20

$$\frac{\tau}{\tau_0} = \frac{v_e}{v} \frac{du^+}{dy^+} \tag{5.12}$$

$$\frac{q}{q_0} = \frac{v_e}{v} \frac{1}{Pr_e} \frac{dT^+}{dy^+} \tag{5.13}$$

$$T_w^+ - T_\infty^+ = \int_0^{\delta^+} Pr_e \frac{du^+}{dy^+} dy^+ \tag{5.14}$$

$$T_w^+ - T_\infty^+ = \frac{(f/2)^{1/2}}{St}, \quad u_\delta^+ = \frac{u_\delta}{u^*} = \left(\frac{2}{f}\right)^{1/2} \tag{5.15}$$

$$\frac{(f_s/2)^{1/2}}{St_s} = \int_0^{y_b^+} (Pr_e - 1)\frac{du^+}{dy^+}dy^+ + \int_0^{\delta^+} \frac{du^+}{dy^+}dy^+ \tag{5.16}$$

$$\frac{\frac{f_s}{2St_s} - 1}{(f_s/2)^{1/2}} = \int_0^{y_b^+} (Pr_e - 1)\frac{du^+}{dy^+}dy^+ \tag{5.17}$$

$$F(Pr) = 12.7\left(Pr^{2/3} - 1\right) \tag{5.18}$$

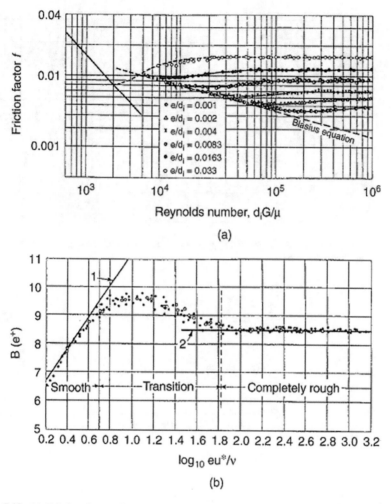

Fig. 5.19 (a) Friction factors for artificially roughened tubes, as measured by Nikuradse. (b) Roughness parameter B for Nikuradse's sand roughness. Curve 1, hydraulically smooth; curve 2, completely rough (Schlichting 1979)

$$\frac{\frac{f_s}{2St_s} - 1}{(f_s/2)^{1/2}} = 12.7\left(Pr^{2/3} - 1\right) \tag{5.19}$$

$$St_s = \frac{f_s/2}{1.0 + 12.7(f_s/2)^{1/2}\left(Pr^{2/3} - 1\right)} \tag{5.20}$$

$$f_s = (1.58 \ln Re_d - 3.28)^{-2} \tag{5.21}$$

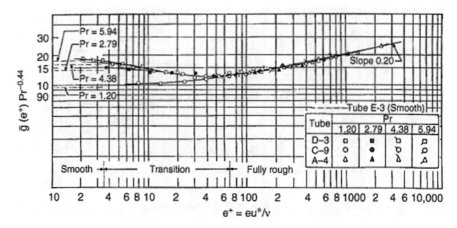

Fig. 5.20 Heat transfer correlation of Dipprey and Sabersky (1963) sand-grain roughness data (Dipprey and Sabersky 1963)

$$\frac{\frac{f}{2St} - 1}{(f/2)^{\frac{1}{2}}} + B(e^+) = \int_0^{e^+} (Pr_e - 1)\frac{du^+}{dy^+} dy^+ \qquad (5.22)$$

$$\bar{g}(e^+)Pr^n = \frac{\frac{f}{2St} - 1}{\sqrt{f/2}} + B(e^+) \qquad (5.23)$$

$$St = \frac{f/2}{1 + \sqrt{f/2}[g(e^+)Pr^n - B(e^+)]} \qquad (5.24)$$

References

Arman B, Rabas TJ (1991) Prediction of the pressure drop in transverse, repeated-rib tubes with numerical modeling. In: Rabas TJ, Chenoweth JM (eds) Fouling and enhancement interactions, ASME HTD, vol 164. ASME, New York, pp 93–99

Arman B, Rabas TJ (1992) Disruption shape effects on the performance of enhanced tubes with the separation and reattachment mechanism. In: Pate MB, Jensen MK (eds) Enhanced heat transfer, ASME Symp. HTD, vol 202. ASME, New York, pp 67–76

Ajeel RK, Salim WI, Hasnan K (2019) Influences of geometrical parameters on the heat transfer characteristics through symmetry trapezoidal-corrugated channel using SiO₂-water nanofluid. Int Commun Heat Mass Transf 101:1–9

Al-Qahtani M, Chen HC, Han JC, Jang YJ (2002) Prediction of flow and heat transfer in rotating two-pass rectangular channels with 45° rib turbulators. ASME J Turbomach 124(2):242–250

Baughn JW, Roby J (1992) Enhanced turbulent heat transfer in circular ducts with transverse ribs. In: Enhanced heat transfer, ASME HTD, vol 202. ASME, New York, pp 9–15

Becker BR, Rivir RB (1989) Computation of the flow field and heat transfer in a rectangular passage with a turbulator, ASME Paper 89-GT-30. ASME, New York

Benodekar RW, Goddard AJH, Gosman AD, Issa RI (1985) Numerical prediction of turbulent flow over surface-mounted ribs. AIAA J 23:359–366

Berger FP, Han K-F (1979) Local mass/heat transfer distribution on surfaces roughened with small square ribs. Int J Heat Mass Transf 22:1645–1656

Chakroun W, Taylor RP (1992) The effects of modestly strong acceleration on heat transfer in the turbulent rough-wall boundary layer. In: Pate MB, Jensen MK (eds) Enhanced heat transfer, SME Symp. HTD, vol 202. ASME, New York, pp 57–66

Chen HC, Patel VC (1988) Near-wall turbulence models for complex flows including separation. AIAA J 26:641–648

Chaube A, Sahoo PK, Solanki SC (2006) Effect of roughness shape on heat transfer and flow friction characteristics of solar air heater with roughened absorber plate. WIT Trans Eng Sci 53:43–51

Cope WG (1945) The friction and heat transmission coefficients of rough pipes. Proc Inst Mech Eng 145:99–105

Dipprey DF, Sabersky RH (1963) Heat and momentum transfer in smooth and rough tubes at various Prandtl numbers. Int J Heat Mass Transf 6:329–353

Durst F, Ponti M, Obi S (1988) Experimental and computational investigation of the two dimensional channel flow over two fences in tandem. J Fluids Eng 110:48–54

Ekkad SV, Han JC (1997) Detailed heat transfer distributions in two-pass square channels with rib turbulators. Int J Heat Mass Transf 40:2525–2537

Esen EB, Obot NT, Rabas TJ (1994a) Enhancement: part I. Heat transfer and pressure drop results for air flow through passages with spirally-shaped roughness. J Enhanc Heat Transf 1:145–156

Esen EB, Obot NT, Rabas TJ (1994b) Enhancement: part II. The role of transition to turbulent flow. J Enhanc Heat Transf 1:157–167

Faramarzi J, Logan E (1991) Reattachment length behind single roughness element in turbulent pipe flow. J Fluids Eng 113:712–714

Farrell P, Wert K, Webb RL (1991) Heat transfer and friction characteristics of turbulator radiator tubes, SAE Technical Paper 910917. SAE International Congress, Detroit, MI

Fodemski TR, Collins MW (1988) Flow and heat transfer simulations for two- and three dimensional smooth and ribbed channels. In: Proceedings of the 2nd U.K national conference on heat transfer, C138/88, University of Stratchlyde, Glasgow, UK

Fujita H, Hajime Y, Nagata C (1986) The numerical prediction of fully developed turbulent flow and heat transfer in a square duct with two roughened facing walls. In: Proceedings of the 8th international heat transfer conference, vol 3, pp 919–924

Gowen RA, Smith JW (1968) Turbulent heat transfer from smooth and rough surfaces. Int J Heat Mass Transf 11:1657–1673

Han JC (1988) Heat transfer and friction characteristics in rectangular channels with rib turbulators. ASME J Heat Transf 110(2):321–328

Han JC, Park JS (1988) Developing heat transfer in rectangle channels with rib turbulators. Int J Heat Mass Transf 31(1):183–195

Han JC, Zhang P (1991) Effect of rib angle orientation on local mass transfer distribution in a three-pass rib-roughened channel. ASME J Turbomach 113(1):123–130

Han JC, Glicksman LR, Rohsenow WM (1978) An investigation of heat transfer and friction for a rib-roughened surfaces. Int J Heat Mass Transf 21(8):1143–1156

Han JC, Park JS, Lei CK (1985) Heat transfer enhancement in channels with turbulence promoters. ASME J Eng Gas Turbine Power 107(3):628–635

Han JC, Ou S, Park JS, Lei CK (1989) Augmented heat transfer in rectangular channels of narrow aspect ratios with rib turbulators. Int J Heat Mass Transf 32(9):1619–1630

Hijikata K, Mori Y (1987) Fundamental study of heat transfer augmentation of tube inside surface by cascade smooth turbulence promoters and its application to energy conversion. Wiinne Stoffiibertrag 21:115–124

Hijikata K, Ishiguro H, Mori Y (1987) Heat transfer augmentation in a pipe flow with smooth cascade turbulence promoters and its application to energy conversion. In: Yang WJ, Mori Y (eds) Heat transfer in high technology and power engineering. Hemisphere, New York, pp 368–397

Hinze JO (1975) Turbulence, 2nd edn. McGraw-Hill, New York

Hung YH, Liou TM, Syang YC (1987) Heat transfer enhancement of turbulent flow in pipes with an external circular rib. In: Jensen MK, Carey VP (eds) Advances in enhanced heat transfer-1987, ASME Symp. HTD, vol 68. ASME, New York, pp 55–64

Iacovides H, Raisee M (1999) Recent progress in the computation of flow and heat transfer in internal cooling passages of turbine blades. Int J Heat Fluid Flow 20:320–328

James CA, Hodge BK, Taylor RP (1994) A validated procedure for the prediction of fully developed Nusselt numbers and friction factors in tubes with two-dimensional rib roughness. J Enhanc Heat Transf 1:287–304

Jia R, Saidi A, Sunden B (2002) Heat transfer enhancement in square ducts with V-shaped ribs of various angles. Proc ASME Turbo Expo 3:469–476

Kanoun M, Baccar M, Mseddi M (2011) Computational analysis of flow and heat transfer in passages with attached and detached rib arrays. J Enhanc Heat Transf 18(2):167–176

Kays WM, Crawford ME (1980) Convective heat transfer. McGraw-Hill, New York, p 174 and 188

Kiml R, Mochizuki S, Murata A (2003) Effects of rib height on heat transfer performance inside a high aspect ratio channel with inclined ribs. J Enhanc Heat Transf 10(4):431–443

Kuwahara H, Takahashi K, Yanagida T, Nakayama W, Hzgimoto S, Oizumi K (1989) Method of producing a heat transfer tube for single-phase flow. U.S. patent 4,794,775, January 3

Lee YO, Ahn J, Lee JS (2008) Effects of dimple depth and Reynolds number on the turbulent heat transfer in a dimpled channel. Prog Comput Fluid Dyn 8:432–438

Lee YO, Ahn J, Kim J, Lee JS (2012) Effect of dimple arrangements on the turbulent heat transfer in a dimpled channel. J Enhanc Heat Transf 19(4):359–367

Li R, He YL, Lei YG, Tao YB, Chu P (2009) A numerical study on fluid flow and heat transfer performance of internally roughened tubes with dimples. J Enhanc Heat Transf 16(3):267–285

Li S, Ghorbani-Tari Z, Xie G, Sundén B (2013) An experimental and numerical study of flow and heat transfer in ribbed channels with large rib pitch-to-height ratios. J Enhanc Heat Transf 20 (4):305–319

Ligrani PM (2013) Heat transfer augmentation technologies for internal cooling of turbine components of gas turbine engines. Int J Rotating Mach 2013. Article ID 275653

Ligrani PM, Oliveira MM, Blaskovich T (2003) Comparison of heat transfer augmentation on techniques. AIAA J 41(3):337–362

Mantle PL (1966) A new type of roughened heat transfer surface selected by flow visualization techniques. In: Proceedings of the 3rd international heat transfer conference, vol 1, pp 45–55

Mendes PRS, Mauricio MHP (1987) Heat transfer, pressure drop, and enhancement characteristics of the turbulent flow through internally ribbed tubes. In: Convective transport, ASME HTD, vol 82, pp 15–22

Nikuradse J (1933) Laws of flow in rough pipes, VD! Forschungsheft, p 361 [English translation, NACA TM-1292 (1965)]

Nunner W (1956) Heat transfer and pressure drop in rough pipes, VDI-Forschungsheft, 455, Ser B 22: 5–39 [English Translation, AERE Lib./Trans 786 (1958)]

Obot NT, Das L, Rabas TJ (2001) Smooth- and enhanced-tube heat transfer and pressure drop. Part II. The role of transition to turbulent flow. In: Shah RK (ed) Proceedings of the third international conference on compact heat exchangers and enhancement technology for the process industries

Olsson CO, Sunden B (1996) Heat transfer and pressure drop characteristics of ten radiator tubes. Int J Heat Mass Transf 39:3211–3220

Olsson CO, Sunden B (1998a) Experimental study of flow and heat transfer in rib-roughened rectangular channels. Exp Therm Fluid Sci 16:349–365

Olsson CO, Sunden B (1998b) Thermal and hydraulic performance of a rectangular duct with multiple V-shaped ribs. J Heat Transf 121:1072–1077

Patankar SV (1990) Numerical prediction of flow and heat transfer in compact heat exchanger passages. In: Shah RK, Kraus AD, Metzger D (eds) Compact heat exchangers. Hemisphere, Washington, DC, pp 191–204

Perng SW, Wu HW (2013) Heat transfer enhancement for turbulent mixed convection in reciprocating channels by various rib installations. J Enhanc Heat Transf 20(2):95–114

Prakash C, Zerkle R (1995) Prediction of turbulent flow and heat transfer in a ribbed rectangular duct with and without rotation. J Turbomachinery 117:255–264

Rau G, Cakan M, Moeller D, Arts T (1998) The effect of periodic ribs on the local aerodynamic and heat transfer performance of a straight cooling channel. ASME J Turbomach 120(2):368–375

Rabas TJ, Arman B (1992) The influence of the Prandtl number of the thermal performance of tubes with the separation and reattachment enhancement mechanism. J Enhanc Heat Transf 1(1):5–21

Rigby DL, Steinthorsson E, Ameri AA (1997) Numerical prediction of heat transfer in a channel with ribs and bleed, ASME Paper 97-GT-431. ASME, New York

Saha AK, Acharya S (2005) Flow and heat transfer in an internally ribbed duct with rotation: an assessment of large eddy simulations and unsteady Reynolds-averaged Navier–Stokes simulations. ASME J Turbomach 127(2):306–320

Schlichting H (1979) Boundary-Layer theory, 7th edn. McGraw-Hill, New York, pp 600–620

Smith JW, Gowen RA (1965) Heat transfer efficiency in rough pipes at high Prandtl number. AIChE J 11:941–943

Song Y, Zheng S, Sunden B, Xie G, Zhou H (2013) Numerical investigation of turbulent heat transfer enhancement in a ribbed channel with upper-downstream-shaped deflectors. J Enhanc Heat Transf 20(5):399–411

Takahashi K, Nakayama W, Kuwahara H (1988) Enhancement of forced convective heat transfer in tubes having three-dimensional spiral ribs. Heat Transf Jpn Res 17(4):12–28

Taylor RP, Hodge BK (1992) Fully-developed heat transfer and friction factor predictions for pipes with 3-dimensional roughness. In: Ebadian MS, Oosthuizen PH (eds) Fundamentals of forced convection heat transfer, ASME Symp. HTD, vol 210, pp 75–84

Taylor RP, Coleman HW, Hodge BK (1984) A discrete element prediction approach for turbulent flow over rough surfaces, Report TFD-84-1. Department of Mechanical Engineering, Mississippi State University

Taylor RP, Scaggs WF, Coleman HW (1988) Measurement and prediction of the effects of nonuniform surface roughness on turbulent flow friction coefficients. J Fluid Eng 110:380–384

Vicente PG, Garcia A, Viedma A (2002a) Experimental study of mixed convection and pressure drop in helically dimpled tubes for laminar and transition flow. Int J Heat Mass Transf 45:5091–5105

Vicente PG, Garcia A, Viedma A (2002b) Heat transfer and pressure drop for low Reynolds turbulent flow in helically dimpled tubes. Int J Heat Mass Transf 45:543–553

Viswanathan AK, Tafti DK (2006) A comparative study of DES and URANS for flow prediction in a two-pass internal cooling duct. ASME J Fluids Eng 128(6):1136–1345

Watkinson AP, Miletti DL, Kubanek GR (1975) Heat transfer and pressure drop of forge-fin tubes in laminar oil flow, ASME paper 75-HT-41, presented at the AIChE-ASME heat transfer conference, San Francisco, Aug 11–13

Webb RL, Eckert ERG, Goldstein RJ (1971) Heat transfer and friction in tubes with repeated rib roughness. Int J Heat Mass Transf 14:601–617

Webb RL, Kim NH (2005) Principles of enhanced heat transfer. Taylor & Francis, New York

Xie G, Zheng S, Zhang W, Sunden B (2013) A numerical study of flow structure and heat transfer in a square channel with ribs combined downstream half-size or same-size ribs. Appl Therm Eng 61:289–300

Zukauskas AA (1972) Heat transfer from tubes in crossflow. In: Hartnett JP, Irvine TF Jr (eds) Advances in heat transfer, vol 8. Academic Press, New York

Chapter 6
2D Roughness, 3D Roughness and Roughness Applications

Good amount of work has been done on 2D and 3D roughness (Figs. 6.1 and 6.2). The roughness elements may be integral to the base surface or they may be in the form of wire coil inserts. A wire coil insert is an attached helical rib roughness. The integral helical rib roughness may be made as single- or multi-start elements, whereas a wire coil insert is a single-start roughness.

The shape of roughness element in a corrugated tube is different from that of a helical integral roughness or a wire coil insert. Figure 6.1 shows flow patterns of Webb et al. (1971) for transverse rib roughness ($\alpha = 90°$) as a function of the dimensionless rib spacing (p/e). Flow separation and reattachment at six to eight rib heights downstream from the rib occur. Edwards and Sheriff (1961) worked with 2D rib on a flat plate, and they observed that the heat transfer coefficient attained its maximum value near the reattachment point. Figures 6.3 and 6.4 show heat transfer coefficient for transverse rib roughness and final friction correlation (Webb et al. 1971).

Sahu and Prasad (2016) presented a review on the performance of solar air heaters with roughened absorber plates. They presented the performance of 35 different roughness geometries used for heat transfer augmentation. They carried out comparison of the performance of different roughness configurations using thermo-hydraulic performance criteria. The roughness elements were mainly categorized into wire ribs, machine ribs and dimple/protrusion ribs.

The wire ribs include transverse ribs (Prasad and Mulick 1983; Prasad and Saini 1988; Gupta et al. 1993; Verma and Prasad 2000; Prasad 2013), inclined ribs (Gupta et al. 1997), expanded wire mesh (Saini and Saini 1997), transverse broken ribs (Sahu and Bhagoria 2005), inclined ribs with gap (Aharwal et al. 2008), combination of transverse and incline ribs (Varun et al. 2008), arc-shaped ribs (Saini and Saini 2008), V-shaped continuous ribs (Momin et al. 2002), V-shaped staggered discrete wire ribs (Muluwork 2000), discrete V-down ribs (Singh et al. 2011), multiple V-ribs (Hans et al. 2010), multi V-shaped rib with gap (Kumar et al. 2013), W-shaped ribs (Lanjewar et al. 2011), discrete W-shaped ribs (Kumar et al. 2008) and angled

© The Author(s), under exclusive license to Springer Nature Switzerland AG 2020   123
S. K. Saha et al., *Insert Devices and Integral Roughness in Heat Transfer Enhancement*, SpringerBriefs in Applied Sciences and Technology,
https://doi.org/10.1007/978-3-030-20776-2_6

Fig. 6.1 Illustrations of
different methods of making
two-dimensional roughness
in a tube (**a**) integral
transverse rib, (**b**)
corrugated transverse rib, (**c**)
integral helical rib, (**d**)
helically corrugated, (**e**)
wire coil insert, (**f**) different
possible profile shapes of
roughness (Ravigururajan
and Bergles 1985)

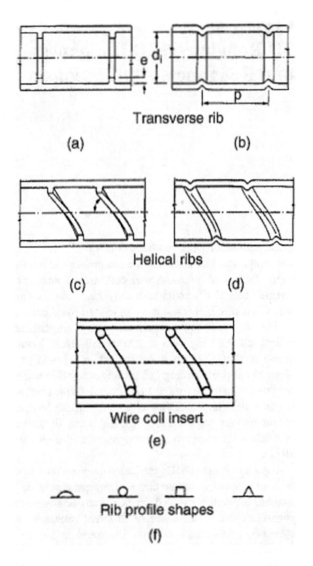

continuous ribs, transverse continuous and broken ribs and V-shaped discrete ribs
(Tanda 2016).

The machine rib configurations are wedge-shaped ribs (Bhagoria et al. 2002),
metal grit ribs (Karmare and Tikekar 2007), U-shaped roughness (Bopche and
Tandale 2009), chamfered ribs (Karwa et al. 1999), rib-grooved combined roughness
(Jaurker et al. 2006) and chamfered rib-grooved roughness (Layek et al. 2006).
Different dimple and protrusion ribs are dimple-shaped roughness (Saini and Verma
2008), protrusion ribs (Bhushan and Singh 2011), protrusion arranged in angular
fashion (Sethi et al. 2012), circular protrusion in the arc shape (Yadav et al. 2013)
and three-sided artificial roughness (Prasad et al. 2014).

Fig. 6.2 Catalogue of flow patterns over transverse rib roughness as a function of rib spacing (Webb et al. 1971)

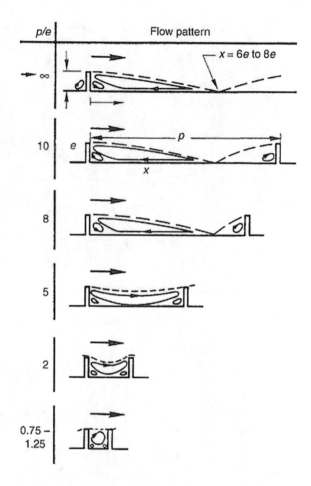

The thermo-hydraulic performance comparison of solar air heaters using different roughness elements has been presented in Fig. 6.5. They concluded that multi V-shaped roughness with gap showed the highest thermo-hydraulic performance parameter of 3.5. Varun et al. (2007), Mittal et al. (2007) and Kumar et al. (2014) have also worked in the same area of heat transfer enhancement in solar air-heaters.

Muluwork (2000), Karwa et al. (1999), Lee and Abdel-Moneim (2001), Slanciauskas (2001), Liou et al. (1993), Rau et al. (1998), Karwa (2003), and Tanda (2004) had reported an experimental investigation of effect of roughened surface on heat transfer characteristics. The heat transfer enhancement results were presented in terms of performance index as described below:

$$\eta = \frac{St}{St_o} \Big/ \left(\frac{f}{f_o}\right)^{1/3} \qquad (6.1)$$

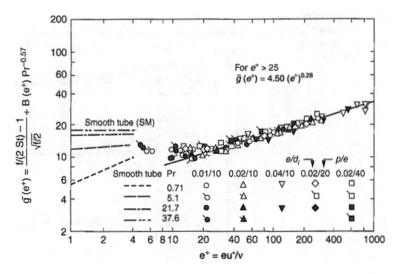

Fig. 6.3 Heat transfer correlation for transverse rib roughness ($a = 90°$) (Webb et al. 1971)

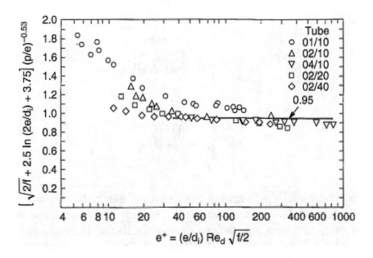

Fig. 6.4 Final friction correlation for repeated-rib tubes (Webb et al. 1971)

Heat transfer and fluid flow characteristics of rib-groove roughened solar air heater ducts had been investigated by Pawar et al. (2009). Figure 6.6 shows the geometry of rib-groove artificial roughness. The performance of solar air heater can be improved by increasing the value of convective heat transfer coefficient between the flowing air and absorber plate. They used integral wedge-shaped rib with and without groove on heated broad wall of the rectangular duct. Laminar sub-layer formed on the absorber plate needs to break down by artificial roughness for increasing the heat transfer rate. The Nusselt number and friction factor of roughened

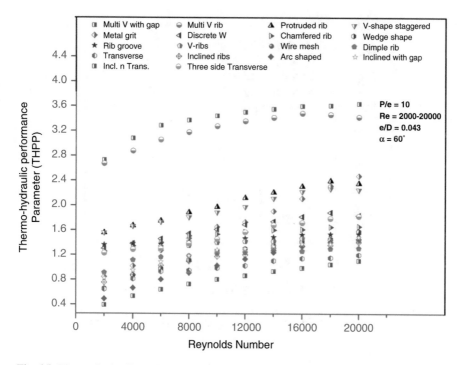

Fig. 6.5 Thermo-hydraulic performance comparison of solar air heaters using different roughness elements (Sahu and Prasad 2016)

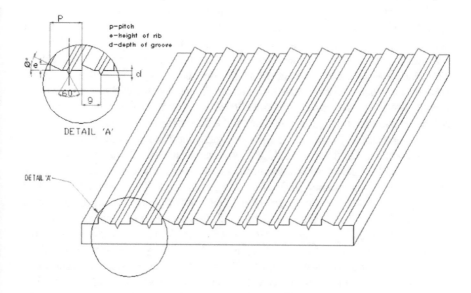

Fig. 6.6 Geometry of the rib-groove artificial roughness (Pawar et al. 2009)

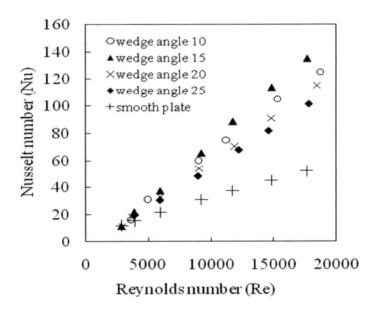

Fig. 6.7 Nusselt number vs. Reynolds number for different values of wedge angle (Pawar et al. 2009)

ducts were determined experimentally and corresponding values were compared with smooth duct in the range of Reynolds number 3000–20,000.

Aspect ratio of rectangular duct was 8. Figure 6.7 shows the variation of Nusselt number with Reynolds numbers for different values of wedge angles, fix values of relative roughness pitch and relative roughness height. At 15° wedge angle of groove was observed optimum for some selected values of Reynolds number for maximum heat transfer enhancement as shown in the Fig. 6.8. They observed that wedge-groove roughened surface enhanced heat transfer more compared to rib roughened surface. It was also observed that both Nusselt number and friction factor increased 2–3 times compared to smooth tube under the operating conditions.

Barba et al. (2002) experimentally investigated thermo-hydraulic performance in the corrugated tube. They used highly viscous Newtonian fluid ethylene glycol whose Prandtl number decreased with the increase in temperature. They worked at moderate Reynolds number ($100 < Re < 800$) and altered it with alteration in mass flow rate. They used the corrugated tube which was manufactured from plain tube. They found lower temperature in corrugated tube than that of plain tube. Nusselt number had oscillating tendency with increasing longitudinal axis. They found that heat transfer enhancement increased significantly with Reynolds number where as Prandtl number did not influence much. The range of heat transfer was in between 4.27 and 16.79. Taking care of actual design they plotted pressure drop against thermal resistance transfer coefficient. They found significant heat transfer augmentation using corrugated tube. Numerically, thermal resistance was 0.2806 °C/W for

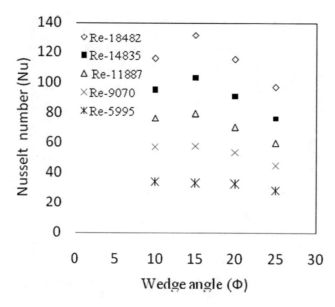

Fig. 6.8 Nusselt number as a function of wedge angle for selected Reynolds number, fixed $e/D = 0.033$ and $p/e = 8$ (Pawar et al. 2009)

plain tube, and it got reduced to 0.01–0.07 °C/W for corrugated tubes. Also, friction factor increased 1.83–2.45 times greater than that of smooth tube.

Dipprey and Sabersky worked with sand-grain roughness. Webb and Eckert (1972) investigated the effect of the p/e and rib cross-sectional shape on the correlation of transverse rib data taken by other investigators. Almeida and Souza-Mendes (1992) measured the length of the thermal entrance region and the local distribution of the mass transfer coefficient between ribs.

Naphon et al. (2006) studied the heat transfer and pressure drop conjugated with the horizontal double pipe of diameter less than 10 mm consisting of helical ribs. They experimented with various flow rates and the inlet temperature for the working fluid, hot and cold water that were entering the section. They concluded from the experimental investigation that heat transfer coefficient increased with the increase in Reynolds number as well as with the increase in helical rib depth that were represented in Fig. 6.9. They established the best fit Nusselt number and friction factor correlation were as

$$Nu_{he} = \frac{h_i d_i}{K} = 44.26 \left(x/d_i \right)^{0.89} \left(p/d_i \right)^{-0.96} (Pe - 1500)^{0.27} Pr^{-0.26} \tag{6.2}$$

$$f_{he} = 7.85 \left(x/d_i \right)^{1.68} \left(p/d_i \right)^{-0.54} Re^{-0.21} \tag{6.3}$$

valid for Reynolds number ranging from 5000 to 25,000 and Prandtl number greater than 3. They found that increase in Nusselt number at higher depth and lower pitch

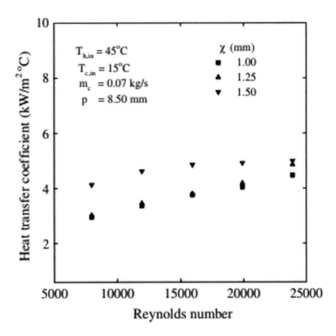

Fig. 6.9 Variation of heat transfer coefficient with Reynolds number for different helical rib depths (Naphon et al. 2006)

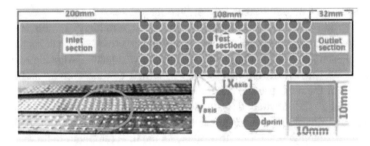

Fig. 6.10 Transverse protruded rib arrangement in the tube (Kumar et al. 2019)

of inserted helical rib was significant than that obtained by lower depth and higher pitch. The causes of pressure drop were drag force, blockage of flow due to reduced area and rotational flow generated by helical ribs. Friction factor was higher with higher rib depth and lower helical rib pitch and vice versa.

Kumar et al. (2019) studied the heat transfer and friction factor characteristics of nanofluid flow in a square cross-sectional channel having protruding ribs. The study was carried out for turbulent flow regime under constant heat flux boundary condition. The Al_2O_3-water nanofluid was used as the working fluid. Figure 6.10 shows the transverse protruded ribs. The effect of ratio of protruded rib height (e_{rib}) to print diameter (d_{print}) of the protruded transverse ribs on overall efficiency has been shown

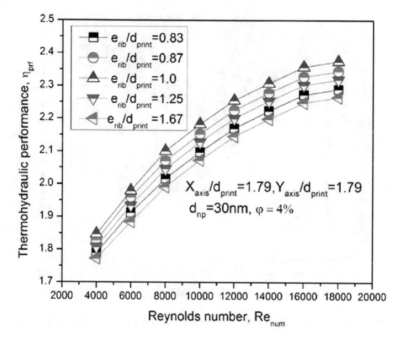

Fig. 6.11 Effect of protruded rib height on thermohydraulic performance (Kumar et al. 2019)

in Fig. 6.11. Also, the variation of thermo-hydraulic performance with ratio of streamwise pitch (X_{axis}) to print diameter and the ratio of spanwise rib pitch (Y_{axis}) and print diameter have been shown in Figs. 6.12 and 6.13, respectively. The extreme values of Nusselt number and friction factor for a fixed geometry of transverse protruded ribs have been shown in Table 6.1. The maximum overall enhancement ratio for the given set of rib parameters has been presented in Table 6.2. They reported that the Nusselt number and friction factor for the rough tube increased with increase in X_{axis}/d_{print} and Y_{axis}/d_{print}. They have concluded from their observations that highest value of overall enhancement ratio obtained was 2.37.

Khalid et al. (2016) used channel with square ribs, Nadankrishnan used combination of nanofluids and staggered dimpled surfaces in microchannel flow. Bianco et al. (2018) numerically studied nanofluid flow through rectangular duct, Ahmed et al. (2015) studied nanofluid flow in triangular channels, Alipour et al. (2017) investigated T-semi-attached ribs along with nanofluids, and Liu et al. (2015) used dimpled surfaces.

Pal and Bhattacharyya (2018) numerically investigated the performance of wall-mounted blunt ribs in a channel flow using nanofluid as working fluid. The Cu-water nanofluid was used. They evaluated the optimum heat transfer enhancement performance by calculating the total entropy generation, surface area to heat transfer augmentation ratio and pressure drop in terms of friction factor. The schematic of the cosine wave-shaped ribs used for the study has been shown in Fig. 6.14. They

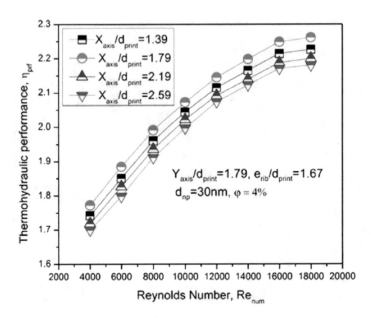

Fig. 6.12 Effect of streamwise pitch on thermohydraulic performance (Kumar et al. 2019)

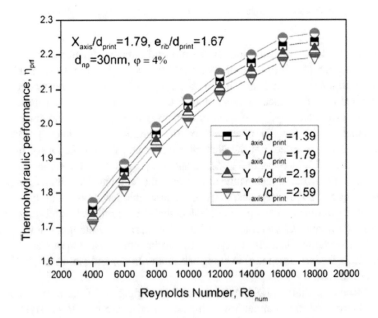

Fig. 6.13 Effect of spanwise pitch on thermohydraulic performance (Kumar et al. 2019)

Table 6.1 Nusselt number and friction factor for a fixed geometry of transverse protruded ribs (Kumar et al. 2019)

Sr. no.	Fixed parameter	Varied parameter	Range of Nu_{rough}	Range of f_{rough}
1	$X_{\text{axis}}/d_{\text{print}} = 1.79$, $Y_{\text{axis}}/d_{\text{print}} = 1.79$, $e_{\text{rib}}/d_{\text{print}} = 1.67$, $d_{\text{np}} = 30$	$\varphi = 1\%$	40–163.8	0.085–0.153
2	$X_{\text{axis}}/d_{\text{print}} = 1.79$, $Y_{\text{axis}}/d_{\text{print}} = 1.79$, $e_{\text{rib}}/d_{\text{print}} = 1.67$, $d_{np} = 30$	$\varphi = 2\%$	43–172.5	0.089–0.159
3	$X_{\text{axis}}/d_{\text{print}} = 1.79$, $Y_{\text{axis}}/d_{\text{print}} = 1.79$, $e_{\text{rib}}/d_{\text{print}} = 1.67$, $d_{np} = 30$	$\varphi = 3\%$	46–180	0.093–0.165
4	$X_{\text{axis}}/d_{\text{print}} = 1.79$, $Y_{\text{axis}}/d_{\text{print}} = 1.79$, $e_{\text{rib}}/d_{\text{print}} = 1.67$, $d_{np} = 30$	$\varphi = 4\%$	49–187	0.101–0.172
5	$X_{\text{axis}}/d_{\text{print}} = 1.79$, $Y_{\text{axis}}/d_{\text{print}} = 1.79$, $e_{\text{rib}}/d_{\text{print}} = 1.67$, $\varphi = 4\%$	$d_{\text{np}} = 30$	49–187	0.101–0.172
6	$X_{\text{axis}}/d_{\text{print}} = 1.79$, $Y_{\text{axis}}/d_{\text{print}} = 1.79$, $e_{\text{rib}}/d_{\text{print}} = 1.67$, $\varphi = 4\%$	$d_{\text{np}} = 35$	46–180	0.098–0.167
7	$X_{\text{axis}}/d_{\text{print}} = 1.79$, $Y_{\text{axis}}/d_{\text{print}} = 1.79$, $e_{\text{rib}}/d_{\text{print}} = 1.67$, $\varphi = 4\%$	$d_{\text{np}} = 40$	43–174.5	0.094–0.162
8	$X_{\text{axis}}/d_{\text{print}} = 1.79$, $Y_{\text{axis}}/d_{\text{print}} = 1.79$, $e_{\text{rib}}/d_{\text{print}} = 1.67$, $\varphi = 4\%$	$d_{\text{np}} = 45$	41–167.8	0.091–0.156
9	$X_{\text{axis}}/d_{\text{print}} = 1.79$, $Y_{\text{axis}}/d_{\text{print}} = 1.79$, $d_{\text{np}} = 30$, $\varphi = 4\%$	$e_{\text{rib}}/d_{\text{print}} = 0.83$	54–196	0.109–0.180
10	$X_{\text{axis}}/d_{\text{print}} = 1.79$, $Y_{\text{axis}}/d_{\text{print}} = 1.79$, $d_{\text{np}} = 30$, $\varphi = 4\%$	$e_{\text{rib}}/d_{\text{print}} = 0.87$	60–209	0.127–0.194
11	$X_{\text{axis}}/d_{\text{print}} = 1.79$, $Y_{\text{axis}}/d_{\text{print}} = 1.79$, $d_{\text{np}} = 30$, $\varphi = 4\%$	$e_{\text{rib}}/d_{\text{print}} = 1.0$	64–217	0.132–0.199
12	$X_{\text{axis}}/d_{\text{print}} = 1.79$, $Y_{\text{axis}}/d_{\text{print}} = 1.79$, $d_{\text{np}} = 30$, $\varphi = 4\%$	$e_{\text{rib}}/d_{\text{print}} = 1.25$	56–202	0.118–0.188
13	$X_{\text{axis}}/d_{\text{print}} = 1.79$, $Y_{\text{axis}}/d_{\text{print}} = 1.79$, $d_{\text{np}} = 30$, $\varphi = 4\%$	$e_{\text{rib}}/d_{\text{print}} = 1.67$	49–187	0.101–0.172
14	$Y_{\text{axis}}/d_{\text{print}} = 1.79$, $e_{\text{rib}}/d_{\text{print}} = 1.67$, $d_{\text{np}} = 30$, $\varphi = 4\%$	$X_{\text{axis}}/d_{\text{print}} = 1.39$	46–180	0.098–0.167
15	$Y_{\text{axis}}/d_{\text{print}} = 1.79$, $e_{\text{rib}}/d_{\text{print}} = 1.67$, $d_{\text{np}} = 30$, $\varphi = 4\%$	$X_{\text{axis}}/d_{\text{print}} = 1.79$	49–187	0.101–0.172
16	$Y_{\text{axis}}/d_{\text{print}} = 1.79$, $e_{\text{rib}}/d_{\text{print}} = 1.67$, $d_{\text{np}} = 30$, $\varphi = 4\%$	$X_{\text{axis}}/d_{\text{print}} = 2.19$	43–174.5	0.091–0.159
17	$Y_{\text{axis}}/d_{\text{print}} = 1.79$, $e_{\text{rib}}/d_{\text{print}} = 1.67$, $d_{\text{np}} = 30$, $\varphi = 4\%$	$X_{\text{axis}}/d_{\text{print}} = 2.59$	41–167.8	0.087–0.152
18	$X_{\text{axis}}/d_{\text{print}} = 1.79$, $e_{\text{rib}}/d_{\text{print}} = 1.67$, $d_{\text{np}} = 30$, $\varphi = 4\%$	$Y_{\text{axis}}/d_{\text{print}} = 1.39$	46–180	0.091–0.164
19	$X_{\text{axis}}/d_{\text{print}} = 1.79$, $e_{\text{rib}}/d_{\text{print}} = 1.67$, $d_{\text{np}} = 30$, $\varphi = 4\%$	$Y_{\text{axis}}/d_{\text{print}} = 1.79$	49–187	0.101–0.172
20	$X_{\text{axis}}/d_{\text{print}} = 1.79$, $e_{\text{rib}}/d_{\text{print}} = 1.67$, $d_{\text{np}} = 30$, $\varphi = 4\%$	$Y_{\text{axis}}/d_{\text{print}} = 2.19$	42.5–172.5	0.091–0.159
21	$X_{\text{axis}}/d_{\text{print}} = 1.79$, $e_{\text{rib}}/d_{\text{print}} = 1.67$, $d_{\text{np}} = 30$, $\varphi = 4\%$	$Y_{\text{axis}}/d_{\text{print}} = 2.59$	40–163.8	0.087–0.152

Table 6.2 The maximum overall enhancement ratio for the given set of rib parameters (Kumar et al. 2019)

Sr. no.	Fixed parameter	Roughness parameter	Optimum data of $\eta_{prf} = \dfrac{Nu_{rough}/Nu_{smooth}}{\left(f_{rough}/f_{smooth}\right)^{0.33}}$
1	$Y_{axis}/d_{print} = 1.79$, $e_{rib}/d_{print} = 1.67$, $d_{np} = 30$, $\varphi = 4\%$	$X_{axis}/$ $d_{print} = 1.79$	2.27
2	$X_{axis}/d_{print} = 1.79$, $e_{rib}/d_{print} = 1.67$, $d_{np} = 30$, $\varphi = 4\%$	$Y_{axis}/$ $d_{print} = 1.79$	2.27
3	$X_{axis}/d_{print} = 1.79$, $Y_{axis}/d_{print} = 1.79$, $d_{np} = 30$, $\varphi = 4\%$	$e_{rib}/d_{print} = 1.0$	2.37

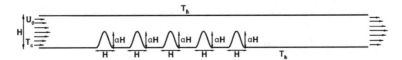

Fig. 6.14 Schematic of the cosine wave-shaped ribs (Pal and Bhattacharyya 2018)

observed flow separation due to the presence of ribs and eddy recirculation downstream of the ribs.

They reported that the effect of ribs was negligible for low Reynolds number ($Re < 50$). However, the improvement in performance has been observed with the addition of nanoparticles to the base fluid. The increase in entropy generation, friction factor and heat transfer rate was observed with increase in the number of ribs. The ratio of heat transfer enhancement and entropy generation was found to be maximum in case of the tube with single rib rather than that with multiple ribs. Guo et al. (2015), Eiamsa-ard and Promvonge (2008, 2009), Chen et al. (2014) and Huang et al. (2013) worked with rib-grooved roughness. Kamali and Binesh (2008), Wang and Chen (2002), Herman and Kang (2002), Wei et al. (2007) and Sui et al. (2010) studied the performance of different insert and roughness configurations.

Cimina et al. (2015) experimentally studied the heat transfer and pressure drop characteristics in a U-bend section. The guide vanes and ribs on the outer wall surface have been used to enhance the rate of heat transfer. They concluded that V-45° downstream-pointing endwall ribs along with guide vanes showed improved performance. Similar works have been carried out by Schüler et al. (2011), Luo and Razinsky (2009), Chen et al. (2011) and Coletti et al. (2013).

Hijikata et al. (1987), Liou and Hwang (1993) and Chandra et al. (2003) investigated the effect of the rib shape on performance. Taslim and Spring (1994) tested ribbed channels with round edges and filleted base ribs, and compared the results with those from sharp-edged ribs. Tanasawa et al. (1983, 1985) tested a perforated form of transverse rib roughness (Fig. 6.15). The friction is smaller because of reduced profile drag. Hwang and Liou (1994) investigated the effect of rib porosity

Fig. 6.15 Transverse rib
promoters investigated
(Tanasawa et al. 1983)

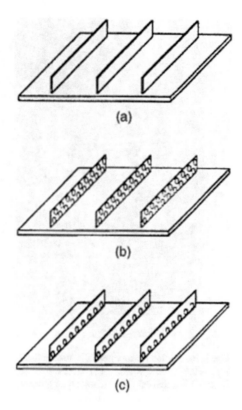

(a)

(b)

(c)

of the hole-type perforated rib geometry. Laser holographic interferometry was
employed to measure the local heat transfer coefficients.

Hwang and Liou (1995) investigated the effect of rib pitch-to-height ratio. Hwang
(1998) investigated the effect of rib porosity of the slit-type rib geometry. The hole-
type perforated ribs provide better thermal performance than the slit-type ribs for the
same pumping power. Gee and Webb (1980), Nakayama et al. (1983), Han et al.
(1978), Withers (1980b), Webb et al. (2000) and Ravigururajan and Bergles (1996)
gave valuable information on all aspects of integral helical rib roughness.

$$\bar{g}(e^+) = 7.68(e^+)^{0.136} \tag{6.4}$$

$$B(e^+) = 1.7 + 2.06 \ln e^+ \tag{6.5}$$

$$\bar{g}(e^+) = 7.68(e^+)^{0.136} \tag{6.6}$$

$$B(e^+) = 1.7 + 2.06 \ln e^+ \tag{6.7}$$

Figure 6.16 shows photos of commercially used enhanced tubes.

Kim (2015) investigated the performance of helically dimpled tubes to enhance
heat transfer for single-phase flow in turbulent flow regime. The transverse ribs,

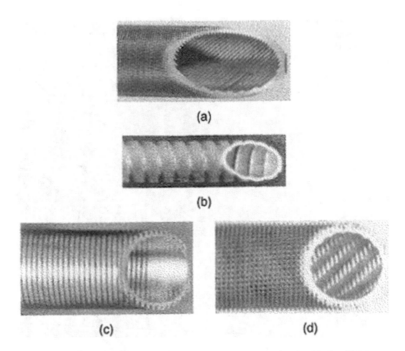

(a)

(b)

(c)   (d)

Fig. 6.16 Photos of commercially used enhanced tubes. (**a**) Wolverine Turbo-C™ (courtesy of Wolverine Tube Division), (**b**) Wieland GEWA-TW™ (courtesy of Wieland-Werke AG), (**c**) Hitachi Thermoexcel-CC™ (courtesy of Hitachi Cable, Ltd), (**d**) Sumitomo Tred-26D™ (courtesy of Sumitomo Light Metal Industries)

Fig. 6.17 Helically dimpled three-dimensional roughnesses (Kim 2015)

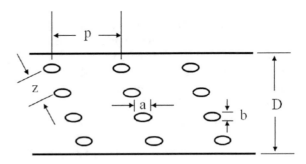

helical ribs, helical corrugations and wire coil inserts are the two-dimensional internal roughness elements. The helically dimpled three-dimensional roughnesses have been shown in Fig. 6.17. The geometrical parameters of the roughness are roughness height (e), axial roughness pitch (p), circumferential roughness pitch (z) and roughness shape (axial distance (a) and transverse distance (b) of the elliptical dimple). The geometrical details of the dimpled tubes have been presented in Table 6.3. Figure 6.18 shows the effect of axial dimple pitch and friction factor (f).

Table 6.3 Geometrical details of the dimpled tubes (Kim 2015)

Tube	D	e	z	p	e/D	z/e	p/e	a	b
Smooth	19.9								
e05z5p3	19.9	0.5	5.0	3.0	0.025	1.0	6.0	2.29	1.70
e05z5p5	19.9	0.5	5.0	5.0	0.025	1.0	10.0	2.29	1.70
e05z5p7	19.9	0.5	5.0	7.0	0.025	14.0	10.0	2.29	1.70
e05z3p3	19.9	0.5	3.0	3.0	0.025	6.0	6.0	2.29	1.70
e05z7p3	19.9	0.5	7.0	3.0	0.025	14.0	6.0	2.29	1.70
e04z5p3	19.9	0.4	5.0	3.0	0.020	12.5	7.5	2.11	1.53
e06z5p3	19.9	0.6	5.0	3.0	0.030	8.3	5.0	2.57	1.93

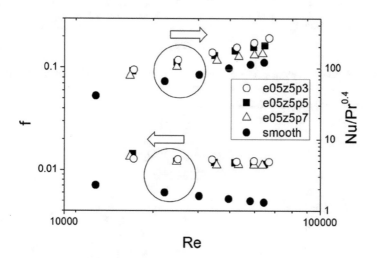

Fig. 6.18 Effect of axial dimple pitch and friction factor (Kim 2015)

They observed an increase in both Nusselt number and friction factor for tubes with roughness over those for smooth tubes. This increase in heat transfer and friction factor has been attributed to the boundary layer disturbance and formation of drag, respectively, due to the presence of surface roughness. They also observed that the increase in roughness height results in increased heat transfer coefficient as long as the roughness height is less than or equal to the boundary layer thickness. When the roughness height exceeds the boundary layer thickness, then the increase in heat transfer coefficient becomes insignificant. The effect of dimple height on friction factor, enhancement ratio and overall performance efficiency has been shown in Figs. 6.19, 6.20 and 6.21, respectively. The correlations for Nusselt number and friction factor have been developed and their coefficients have been summarized in Table 6.4. The dimple having 0.5 mm height, 3 mm axial pitch and 5 mm circumferential pitch has been proposed as the optimum configuration.

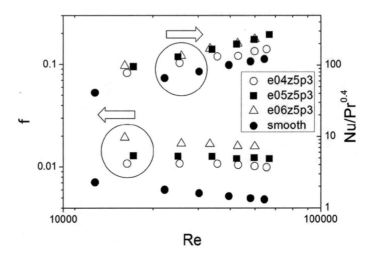

Fig. 6.19 Effect of dimple height on the heat transfer coefficient and the friction factor (Kim 2015)

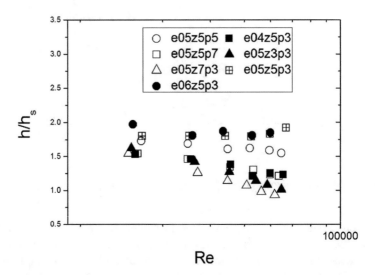

Fig. 6.20 Heat transfer enhancement factors of the dimpled tubes (Kim 2015)

Thianpong et al. (2009), Suresh et al. (2001) and Kumbhar and Sane (2015) have also studied the heat transfer enhancement performance of three-dimensional roughness elements.

Khalid et al. (2016) numerically studied the heat transfer augmentation performance of square ribbed channel having varying rib–pitch ratio. The heat transfer enhancement for cooling blades of advanced gas turbine application has been considered. The rib configurations used for the study have been shown in

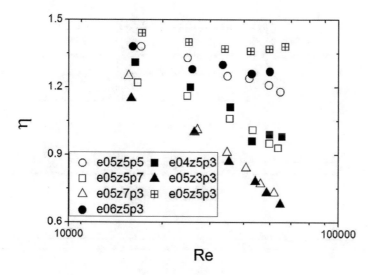

Fig. 6.21 Thermal efficiencies of the dimpled tubes (Kim 2015)

Table 6.4 Correlations for Nusselt number and friction factor have been developed and their coefficients (Kim 2015)

Tube	$Nu\ Pr^{-0.4}$		f	
	a	b	c	d
e05z5p3	0.0213	0.851	0.0243	−0.0563
e05z5p5	0.0946	0.699	0.0833	−0.183
e05z5p7	0.269	0.582	0.0609	−0.157
e05z3p3	1.45	0.415	0.0411	−0.0819
e05z7p3	2.13	0.369	0.0577	−0.155
e04z5p3	0.267	0.583	0.0193	−0.0581
e06z5p3	0.0613	0.752	0.106	−0.175

Fig. 6.22. The variation of Nusselt number and friction factor with Reynolds number for four different configurations has been shown in Fig. 6.23a, b, respectively.

Han et al. (1978, 2000), Murata and Mochizuki (2001), Taslim et al. (2001), Chandra et al. (2003), Ligrani et al. (2003), Eiamsa-ard and Promvonge (2009), Han and Huh (2010), Chyu and Siw (2013), Gupta et al. (2012) and Wright and Han (2014) have also investigated mechanisms for cooling the gas turbines. More information on rib configuration for heat transfer augmentation in flow through channels can be obtained from Li et al. (2012), Xie et al. (2013), Saidi and Sunden (2000), Peng and Peterson (1996), Ooi et al. (2002), Withada and Boonloi (2014), Jia et al. (2005), Withada et al. (2011), Wang and Sunden (2005), Wongcharee et al. (2011) and Kong et al. (2016).

The enhancement mechanism of wall-attached roughness like wire coil inserts is the same as that of helical rib roughness. However, wire coil inserts are single-start roughness, and they have larger wire pitch (and p/e) than the multi-start helical rib roughness. The Sethumadhavan and Raja Rao (1983) correlation is very useful for

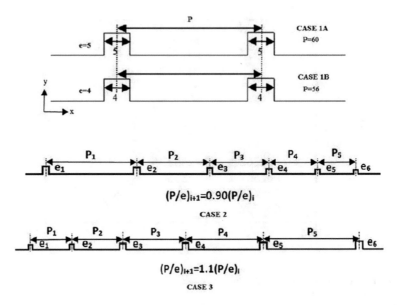

Fig. 6.22 Rib configurations used for the study (Khalid et al. 2016)

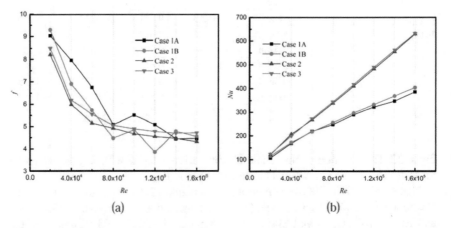

Fig. 6.23 Variation of (**a**) Nusselt number and (**b**) friction factor with Reynolds number for four different configurations (Khalid et al. 2016)

turbulent flow wire coil inserts application. Additionally, Kumar and Jude (1970), Prasad and Saini (1988) and Zhang et al. (1991) are preferred for useful correlations.

$$B(e^+) = 7.0(\tan\alpha)^{-0.18}(e^+)^{0.13} \tag{6.8}$$

$$\bar{g}(e^+) = 8.6(\tan\alpha)^{-0.18}Pr^{-0.55}(e^+)^{0.13} \tag{6.9}$$

Table 6.5 Data range for corrugated tubes (Webb and Kim 2005)

Reference	Fluid	No.	e/d_i	p/e	α
Withers (1980a)	Water	14	0.016–0.043	11–24	79–85
Mehta and Raja Rao (1988)	Water	11	0.008–0.097	4.5–49	69–84
Sethumadavan and Raja Rao (1986)	Water glycerin	5	0.012–0.030	13–45	65
Raja Rao (1988)	Water power law	12	0.020–0.060	4.6–30	65–82
Li et al. (1982)	Water	20	0.008–0.069	7.7–29	41–85
Dong et al. (2001)	Water and oil	4	0.020–0.040	14.5–31.2	79–82

$$Nu_d = 0.175 \left(\frac{p}{d_i}\right)^{-0.35} Re_d^{0.7} Pr^{1/3} \tag{6.10}$$

$$Nu_d = 0.235 Re_d^{0.716} \left(\frac{e}{d_i}\right)^{0.372} \left(\frac{p}{d_i}\right)^{0.171} \tag{6.11}$$

$$f = 62.36 \ln Re_d^{-2.78} \left(\frac{e}{d_i}\right)^{0.816} \left(\frac{p}{d_i}\right)^{-0.689} \; ; \quad 6 \leq Re_a \times 10^{-3} \langle 15 \tag{6.12}$$

$$f = 5.153 \ln Re_d^{-1.08} \left(\frac{e}{d_i}\right)^{0.796} \left(\frac{p}{d_i}\right)^{-0.707} \; ; \quad 15 \leq Re_a \times 10^{-3} \langle 100 \tag{6.13}$$

Wire coil inserts may be of circular or rectangular cross section.

Data on single- and multi-start corrugated tubes have been given by Newson and Hodgson (1973), Sethumadhavan and Raja Rao (1986), Li et al. (2012), Dong et al. (2001), Webb et al. (1971), Withers (1980a), Mehta and Raja Rao (1979, 1988), Raja Rao (1988), Ravigururajan and Bergles (1985), Rabas et al. (1988). Tables 6.5, 6.6, 6.7 and 6.8 give some useful information on corrugated tube roughness. Churchill (1973) correlation has been modified by Withers (1980a) for commercially rough tubes.

$$\sqrt{\frac{2}{f}} = -2.46 \ln \left[\frac{e}{d_i} + \left(\frac{7}{Re_d}\right)^{0.9}\right] \tag{6.14}$$

$$\sqrt{\frac{2}{f}} = -2.46 \ln \left[r + \left(\frac{7}{Re_d}\right)^{m}\right] \tag{6.15}$$

Table 6.6 Correlations for g and $B(e^+)$ (Webb and Kim 2005)

Reference	$\bar{g}(e^+)Pr^n$	$B(e^+)$
Withers (1980)	$\dfrac{4.95}{(\tan\alpha)^{0.33}(e^+)^{0.127}Pr^{0.5}}$	Does not use $B(e^+)$
Mehta and Raja Rao (1988)	$\dfrac{7.92}{(\tan\alpha)^{0.15}(e^+)^{0.11}Pr^{0.55}}$	$0.465(p/e)^{0.53}(\ln e^+ + 0.25)$
Raja Rao (1988)	$\dfrac{6.06}{(\tan\alpha)^{0.15}(e^+)^{0.13}Pr^{0.55}}$	$0.465(p/e)^{0.53}(\ln e^+ + 0.25)(n')^{2.5}$
Sethumadavan and Raja Rao (1986)	$8.6(e^+)^{0.13}Pr^{0.55}$	$0.40(e^2/pd_i)^{-1/3}(e^+)^{0.164}$
Dong et al. (2001)	$7.33(e^+)^{0.17}(\alpha/50)^{-0.16}Pr^{0.548}$	$0.466(e^2/pd_i)^{-0.317}(e^+)^{0.169}(\alpha/50)^{-0.16}$

Table 6.7 Predicted g function for $\alpha = 70°$ and $90°$ (Webb and Kim 2005)

	$[\bar{g}(e^+)(\tan\alpha)^n]$for		
Correlation	$\alpha = 70°$	$\alpha = 85°$	Comments
Withers (1980a)	10.38	16.64	Water
Mehta and Raja Rao (1988)	13.11	16.25	Water
Raja Rao (1988)	10.72	13.28	Water and power law
Sethumadavan and Raja Rao (1986)	13.07	NA	Water and water/glycol

Table 6.8 Tubes tested (Withers 1980a)

Tube no.	d_i (in.)	e/d_i	p/e	α (deg)	m	r
2300	0.892	0.016	20.10	83.03	0.71	0.0001
2100	0.869	0.016	18.76	84.58	0.68	0.00094
2200	0.862	0.020	14.45	84.73	0.564	0
2	0.805	0.024	19.46	81.70	0.649	0
1	0.805	0.024	19.46	81.67	0.622	0
LPD	All	0.025	20.00	81.00	0.61	0.00088
9	1.175	0.030	11.31	83.88	0.50	0.0035
1100	0.547	0.031	14.70	81.72	0.65	0.0023
7	0.922	0.033	12.14	82.73	0.47	0.0015
20	0.855	0.036	12.42	81.85	0.445	0
6	0.921	0.038	11.00	81.53	0.45	0.0073
15	1.152	0.040	10.98	82.06	0.48	0.00995
MHT	All	0.040	12	81	0.44	0.00595
14	0.866	0.043	14.20	78.98	0.52	0.0006
33	0.863	0.047	12.28	79.52	0.47	0.0081
5	0.682	0.052	17.45	73.83	0.51	0.0027

Fig. 6.24 Boundary layer development over the surface of trapezoidal configuration (Cernecky et al. 2015)

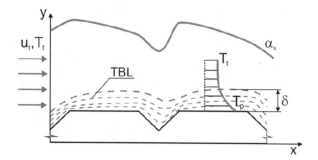

$$r = 0.17 \left(\frac{e}{d_i}\right)^{-1/3} \left(\frac{p}{d_i}\right)^{0.03} \left(\frac{\alpha}{90}\right)^{-0.29} \quad \text{and}$$

$$m = 0.0086 + 0.033 \left(\frac{e}{d_i}\right) + 0.005 \left(\frac{p}{d_i}\right) + 0.0085 \left(\frac{\alpha}{90}\right) \tag{6.16}$$

Naphon (2007) studied triangular grooved channels, Elshafei et al. (2010) used rectangular grooves, Vyas et al. (2010) investigated the sinusoidal corrugated tubes for heat transfer enhancement.

The trapezoidal shaped corrugations for heat transfer augmentation have been considered by Cernecky et al. (2015). They used holographic interferometry (non-contact visualization method) in order to visualize the temperature profiles around the area of heated surfaces. The fluid flow (air flow) was considered between two heated surfaces. The boundary layer development over the surface of trapezoidal configuration has been presented in Fig. 6.24.

The development of boundary layer results in very low heat transfer coefficients. Thus, the mechanism of breaking or disturbing the boundary layer using trapezoidal corrugations has been considered to enhance the heat transfer rate. Tables 6.9 and 6.10 present the local and mean heat transfer parameters, namely heat transfer coefficient and Nusselt number at different axial positions on the lower heated surface and upper heated surface, respectively. Figure 6.25 illustrates the mean Nusselt number for different heat exchangers. Also, the j/f ratio has been shown in Fig. 6.26. They concluded that the highest mean Nusselt number has been observed for the heat exchanger with 'C' configuration on the lower heated surface. They have observed that the addition of directional tubing resulted in increased heat transfer rates at the cost of pressure drop penalty.

Cope (1945) was the pioneer for testing 3D roughness. Subsequently, Gowen and Smith (1968) embossed a 3D roughness pattern on a thin sheet. The sheet was rolled into a turbular form and the axial seam was soldered. Nowadays, with the maturing of manufacturing process, practical 3D roughness configurations are available. Takahashi et al. (1988), Fenner and Ragi (1979) and McLain (1975) (Table 6.11) provide valuable information on 3D roughness. Modern high-speed tube welding techniques make 3D roughness a very practical and useful concept for commercial production. Tubes having inside roughness are now being routinely used.

Table 6.9 Local and mean heat transfer parameters in investigated areas of heat exchange systems A1, A2, B1, B2 and C—lower heated surfaces (Cernecky et al. 2015)

Section	α_x (W/m² K)				
	A1	A2	B1	B2	C
X_1	13.22	4.83	9.98	8.75	16.44
X_2	12.75	6.47	8.95	12.23	18.41
X_3	2.61	2.77	2.68	4.27	3.03
X_4	13.99	6.70	14.05	12.34	18.15
X_5	12.26	5.33	14.82	10.48	20.62
α_m (W/m² K)	**10.97**	**5.22**	**10.10**	**9.61**	**15.33**
Nu_m (−)	**25.41**	**12.09**	**20.28**	**19.29**	**30.78**

Table 6.10 Local and mean heat transfer parameters in investigated areas of heat exchange systems A1, A2, B1, B2 and C—upper heated surfaces (Cernecky et al. 2015)

Section	α_x (W/m² K)				
	A1	A2	B1	B2	C
$X_{1'}$	6.09	9.92	10.37	13.00	3.18
$X_{2'}$	5.83	11.79	11.28	16.69	17.47
$X_{3'}$	5.53	12.48	2.69	4.14	15.91
$X_{4'}$	5.98	12.51	17.02	18.79	18.08
$X_{5'}$	5.75	10.63	16.46	14.88	3.02
$\alpha_{m'}$ (W/m² K)	**5.84**	**11.47**	**11.56**	**13.50**	**11.53**
$Nu_{m'}$ (−)	**13.53**	**26.57**	**23.21**	**27.10**	**23.15**

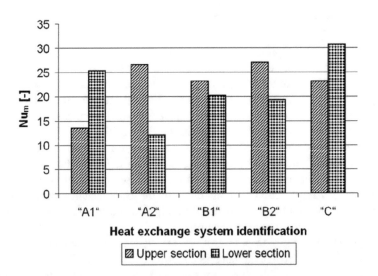

Fig. 6.25 Mean Nusselt number for different heat exchangers (Cernecky et al. 2015)

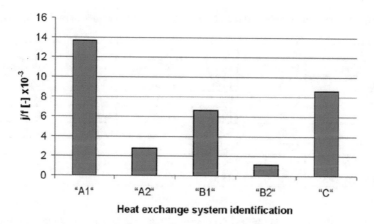

Fig. 6.26 Comparison of heat exchange systems in terms of heat transfer and pressure losses (Cernecky et al. 2015)

Table 6.11 Case VG-1 PEC for Takahashi et al. (1988) tests (copper tubes, $d_i = 13.9$ mm) (Webb and Kim 2005)

Geo	Type	e/d_i	α (deg)	p/e	A/A_s	G/G_s
B3	2-D	0.021	80	13.1	0.79	0.71
A6	3-D	0.022	30	12.6	0.78	0.80
A8	3-D	0.036	30	7.60	0.73	0.72

Wolverine, Turbo-Chil and Korodense tubes, Hitachi Thermoexcel-CC and GEWA-TW are the examples.

$$\frac{hd_i}{k} = 0.06Re_d^{0.8}Pr^{0.4} \tag{6.17}$$

$$f = 0.198Re_d^{-0.267} \tag{6.18}$$

Thors (2004) and Thors et al. (1997) (Tables 6.12 and 6.13) showed some useful data. Rabas et al. (1993) and Kuwahara et al. (1989) discussed the ways to make 3D roughness patterns. Vicente et al. (2002) tested helically dimpled tubes and derived roughness functions

$$B(e^+) = 0.839\left(\frac{e}{d_i}\right)^{-0.85}\left(\frac{d_i^2}{pz}\right)^{-0.18}(e^+)^{0.11} \tag{6.19}$$

$$\bar{g}(e^+)Pr^n = 4.87Pr^{0.60}(e^+)^{0.24} \tag{6.20}$$

Data on transverse and helical rib roughness flow in annuli and rod bundles may be obtained from Sheriff and Gumley (1966), Wilkie (1966), Webb and Eckert (1972),

Table 6.12 Geometric details of the Turbo-B series tubes (dimensions in mm) (Webb and Kim 2005)

Tube no.	I	II	III	IV	V
Product name	Turbo-Chil	Turbo-B	Turbo-BII	Turbo-BIII	Turbo-BIII LPD
Fins/m	1575	1575	1969	2362	2362
Fin height	1.32	0.61	0.69	0.55	0.55
ID	14.55	16.05	16.05	16.38	16.38
Rib height	0.381	0.559	0.381	0.406	0.368
Rib pitch	4.27	2.36	1.07	1.31	1.31
No. of starts	10	30	38	34	34
Helix angle (deg)	46.5	33.5	49	49	49

Table 6.13 Performance comparison of commercially enhanced tubes ($d_0 = 19$ mm, $Re = 25,000$, $Pr = 10.4$) (Webb and Kim 2005)

Tube	d_i	e/d_i	n_s	p/e	α	h/h_p	f/f_p	η
GEWA-TW™	15.3	0.016	1	5.3	89	1.40	1.40	1.00
Thermoexcel-CC™	14.97	0.025	1	46.7	73	1.59	1.90	0.84
GEWA-SC™	15.02	0.035	25	2.67	30	1.87	1.65	1.13
Korodense (LPD)™	17.63	0.025	1	20.3	81	1.89	2.26	0.84
Turbo-Chil™	14.60	0.026	10	11.1	47	1.98	1.83	1.08
Korodense (MHT)™	17.63	0.04	1	12.0	81	2.50	4.63	0.54
Tred-26D™	14.45	0.024	10	7.63	45	2.24	1.88	1.19
Turbo-B™	16.05	0.028	30	1.94	35	2.34	2.14	1.09
Turbo-BIII LPD™	16.38	0.022	34	3.56	49	2.40	1.98	1.21
Turbo-BIII™	16.38	0.025	34	3.22	49	2.54	2.30	1.10
Tred-19D™	14.45	0.024	10	7.63	57	2.55	1.76	1.45
A8 (table 9.9)	13.5	0.036	2	7.6	30	3.75	3.35	1.11

Wilkie et al. (1967), White and Wilkie (1970) and Williams et al. (1970). Velocity profile data may be obtained from Lewis (1974), Maubach (1972), Meyer (1980) and Hudina (1979).

Xie et al. (2013) numerically studied heat transfer augmentation in ribbed channels with square cross-sectioned ribs having differently positioned deflectors. Kanoun et al. (2011), Al-Qahtani et al. (2002), Saha and Acharya (2005) and Viswanathan and Tafti (2006) carried out similar numerical studies.

The effect of rib installations in reciprocating channels for heat transfer augmentation was presented by Perng and Wu (2013). They carried out a numerical analysis and studied the effect of rib–pitch, rib–land and rib–blockage ratios on heat transfer in turbulent flow. They observed 94.44–184.65% increase in average time-mean Nusselt number in the ribbed channel over that in a smooth/plain channel which is subjected to reciprocating motion. They attributed this increase in heat transfer to the combined effect of buoyancy momentum of fluid at the inlet and the reciprocating forces. They concluded that the heat transfer from the wall to the turbulent flow was observed to increase with the increase in rib–land and rib–blockage ratios, and the reverse trend was observed for the increase in rib–pitch ratio.

Han (1988) gave the formula for calculating composite friction factor of a channel having a number of rough and smooth walls.

$$\bar{f} = \frac{L_s f_s + L_r f}{L_s + L_r} \tag{6.21}$$

Han (1984, 1988), Taslim and Spring (1988), Han et al. (1989, 1991, 1993, 2000) and Han and Zhang (1992) worked with rectangular channels. Hishida and Takase (1987), Kukreja et al. (1993), Ekkad and Han (1997), and Ekkad et al. (1998) are more relevant literature on the research topic. Mahmood et al. (2001a, b), Ligrani et al. (2003), Mahmood and Ligrani (2002) and Burges et al. (2003) investigated the flow through channels having concave dimpled geometry (Figs. 6.27 and 6.28), which provide moderately high heat transfer with relatively low pressure drop increase since the dimples do not protrude into the flow and there is almost no form drag.

The flow visualization study showed several longitudinal vortex pairs shed from the edge of the dimples, and these vortex structures enhance heat transfer near the downstream rims of the dimples. Both heat transfer and pressure drop increase with the increase in the ratio of the dimple depth to dimple diameter. Mahmood and Ligrani (2002) observed similar increase with the decrease in the ratio of channel height to dimple diameter. Addition of protrusions to the top wall improves the bottom dimpled surface heat transfer significantly (Ligrani et al. 2001; Mahmood et al. 2001b). Chyu et al. (1997) used dimples on two opposite channel surfaces. Ligrani et al. (2003) compared the performance of the dimpled surface with that of a ribbed surface.

Studies of knurled roughness on the outside of tubes have been performed with air or oil. Groehn and Scholz (1976), Achenbach (1977) and Zukauskas and Ulinskas (1983, 1988) gave a correlation for Nusselt number.

$$Nu_d = 0.5 Re_d^{0.65} Pr^{0.36} \left(\frac{S_t}{S_l}\right)^{0.2} \left(\frac{e}{d_o}\right)^{0.1} \left(\frac{Pr}{Pr_w}\right)^{0.25} \tag{6.22}$$

Knurled roughness is not good for shell-side enhancement in tube banks; there low integral finned tubes are used.

Kang (2001), Xie et al. (2013), Herman and Kang (2001, 2002) and Ortiz et al. (2008) investigated the performance of grooved channels. The grooves provided in channels aid in disturbing the boundary layer in order to achieve the objective of heat transfer augmentation. However, the space between two successive grooves experiences flow stagnation which cannot be completely eliminated. However, flow deflectors used in combination with grooves result in reducing the flow stagnation. This reduction can be achieved only at the cost of pressure drop penalty. The

Fig. 6.27 Concave dimple geometry: (**a**) entire dimpled test surface, (**b**) individual dimple geometry. All dimensions are given in mm (Mahmood et al. 2001a)

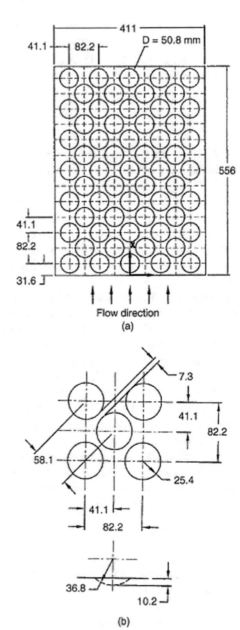

deflector shape and the appropriate location of the deflectors dictate the heat transfer enhancement.

Lee et al. (2012) studied the effect of dimple arrangement on heat transfer enhancement in a dimpled channel turbulent flow. They concluded that the staggered array of dimples showed better enhancement performance than that of in-line array

Fig. 6.28 Sketch of three-dimensional flow structure and flow visualization photographs taken at different planes for $Re = 1250$ and $H/d = 0.5$ (Mahmood et al. 2001a)

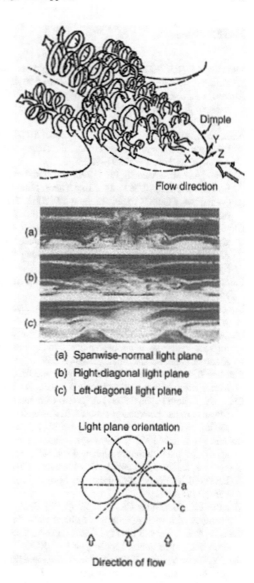

(a) Spanwise-normal light plane

(b) Right-diagonal light plane

(c) Left-diagonal light plane

Light plane orientation

Direction of flow

of dimples. In the dimpled channels, the flow separation and reattachment takes place inside the dimple. Thus, heat transfer augmentation can be observed around the reattachment point. The portion of the reattached flow going out of the dimple increases the fluid mixing by interacting with the flow passing over the dimple. The vortex pairs which are generated at the downstream of the dimple flow along the direction of the dimple diagonal are responsible for the enhancement of heat transfer. Based on large eddy simulation, Lee et al. (2008) explained that the vortices which are generated at the shear layer, as a consequence of separation of the flow, play a vital role in heat transfer augmentation.

References

Achenbach E (1977) The effect of surface roughness on the heat transfer from a circular cylinder to the cross flow of air. Int J Heat Mass Transf 20:359–369

Aharwal KR, Gandhi BK, Saini JS (2008) Experimental investigation on heat-transfer enhancement due to a gap in an inclined continuous rib arrangement in a rectangular duct of solar air heater. Renew Energy 33:585–596

Ahmed HE, Ahmed MI, Yusoff MZ, Hawlader MNA, Al-Ani H (2015) Experimental study of heat transfer augmentation in non-circular duct using combined Nanofluids and vortex generator. Int J Heat Mass Transf 90:1197–1206

Almeida IA, Souza-Mendes PR (1992) Local and average transport coefficients for the turbulent flow in internally ribbed tubes. Exp Therm Fluid Sci 5:513–523

Al-Qahtani M, Chen HC, Han JC, Jang YJ (2002) Prediction of flow and heat transfer in rotating two-pass rectangular channels with 45∘ rib turbulators. ASME J Turbomach 124(2):242–250

Alipour H, Karimipour A, Safaei MR, Semiromi DT, Akbari OA (2017) Influence of t-semi attached rib on turbulent flow and heat transfer parameters of a silver-water nanofluid with different volume fractions in a three-dimensional trapezoidal microchannel. Phys E Low Dimen Syst Nanostruct 88:60–76

Barba A, Rainieri S, Spiga M (2002) Heat transfer enhancement in a corrugated tube. Int Commun Heat Mass Transf 29(3):313–322

Burgess NK, Oliveira MM, Ligrani PM (2003) Nusselt number behavior on deep dimpled surfaces within a channel. J Heat Transf 125:11–18

Bhagoria JL, Saini JS, Solanki SC (2002) Heat transfer coefficient and friction factor correlations for rectangular solar air heater duct having transverse wedge shaped rib roughness on the absorber plate. Renew Energy 25:341–369

Bhushan B, Singh R (2011) Nusselt number and friction factor correlations for solar air heater duct having artificially roughened absorber plate. Sol Energy 85:1109–1118

Bianco V, Scarpa F, Tagliafico LA (2018) Computational fluid dynamics modeling of developing forced laminar convection flow of Al_2O_3–water Nanofluid in a two-dimensional rectangular section channel. J Enhanc Heat Transf 25(4–5):387–398

Bopche SB, Tandale MS (2009) Experimental investigations on heat transfer and frictional characteristics of a turbulator roughened solar air heater duct. Int J Heat Mass Transf 52:2834–2848

Cernecky J, Koniar J, Ohanka L, Brodnianska Z (2015) Temperature field and heat transfer in low REYNOLDS flows inside trapezoidal-profiled corrugated-plate channels. J Enhanc Heat Transf 22(4):329–343

Chandra PR, Alexander CR, Han JC (2003) Heat transfer and friction behaviors in rectangular channels with varying number of ribbed walls. Int J Heat Mass Transf 46:481–495

Chen W, Ren J, Jiang H (2011) Effect of turning vane configurations on heat transfer and pressure drop in a ribbed internal cooling system. ASMEJ Turbomach 133(4):041012

Chen Y, Chew Y, Khoo B (2014) Heat transfer and flow structure on periodically dimple protrusion patterned walls in turbulent channel flow. Int J Heat Mass Transf 78:871–882

Churchill SW (1973) Empirical expressions for the shear stress in turbulent flow in commercial pipe. AiChE J 19:375–376

Chyu MK, Yu Y, Ding H, Downs JP, Soechting FO (1997) Concavity enhanced heat transfer in an internal cooling passage, ASME Paper 97-GT-437. ASME, New York

Chyu MK, Siw SC (2013) Recent advances of internal cooling techniques for gas turbine air foils. ASME J Therm Sci Eng Appl 5(021008):1–12

Cimina S, Wang C, Wang L, Niro A, Sunden B (2015) Experimental study of pressure drop and heat transfer in a u-bend channel with various guide vanes and ribs. J Enhanc Heat Transf 22 (1):29–45

Coletti F, Verstraete T, Bulle J, Van derWielen T, Van den Berge N, Arts T (2013) Optimization of a U-bend for minimal pressure loss in internal cooling channels—part II: experimental validation. ASMEJ Turbomach 135(5):051016

Cope WG (1945) The friction and heat transmission coefficients of rough pipes. Proc Inst Mech Eng 145:99–105

Dong Y, Huixiong L, Tingkuan C (2001) Pressure drop, heat transfer and performance of singlephase turbulent flow in spirally corrugated tubes. Exp Therm Fluid Sci 24:131–138

Edwards FJ, Sheriff N (1961) The heat transfer and friction characteristics of forced convection air flow over a particular type of rough surface. In: International developments in heat transfer. ASME, New York, pp 415–426

Eiamsa-ard S, Promvonge P (2008) Numerical study on heat transfer of turbulent channel flow over periodic grooves. Int Commun Heat Mass Transf 35(7):844–852

Eiamsa-ard S, Promvonge P (2009) Thermal characteristics of turbulent rib-grooved channel flows. Int Commun Heat Mass Transf 36(7):705–711

Ekkad SV, Han JC (1997) Detailed heat transfer distributions in two-pass square channels with rib turbulators. Int J Heat Mass Transf 40:2525–2537

Ekkad SV, Huang Y, Han JC (1998) Detailed heat transfer distributions in two-pass smooth and turbulated square channels with bleed holes. Int J Heat Mass Transf 41:3781–3791

Elshafei EAM, Awad MM, El-Negiry E, Ali AG (2010) Heat transfer and pressure drop in corrugated channels. Energy 35(1):101–110

Fenner GW, Ragi E (1979) Enhanced tube inner surface heat transfer device and method. U.S. patent 4,154,291, May 15

Gee DL, Webb RL (1980) Forced convection heat transfer in helically rib-roughened tubes. Int J Heat Mass Transf 23:1127–1136

Gowen RA, Smith JW (1968) Turbulent heat transfer from smooth and rough surfaces. Int J Heat Mass Transf 11:1657–1673

Groehn HG, Scholz F (1976) Heat transfer and pressure drop of in-line tube banks with artificial roughness. In: Heat and mass transfer sourcebook: fifth all-union conference, Minsk, Scripta, Washington, DC, pp 21–24

Guo L, Xu H, Gong L (2015) Influence of wall roughness models on fluid flow and heat transfer in microchannels. Appl Therm Eng 84:399–408

Gupta D, Solanki SC, Saini JS (1993) Heat and fluid flow in rectangular solar air heater ducts having transverse rib roughness on absorber plates. Sol Energy 51(1):31–37

Gupta D, Solanki SC, Saini JS (1997) Thermohydraulic performance of solar air heaters with roughened absorber plates. Sol Energy 61:33–42

Gupta S, Chaube A, Verma P (2012) Review on heat transfer augmentation techniques: application in gas turbine blade internal cooling. J Eng Sci Technol Rev 5:57–62

Han JC (1984) Heat transfer and friction in channels with two opposite rib-roughened walls. J Heat Transf 106:774–781

Han JC (1988) Heat transfer and friction characteristics in rectangular channels with rib turbulators. J Heat Transf 110:321–328

Han JC, Zhang YM (1992) High performance heat transfer ducts with parallel broken and V-shaped broken ribs. Int J Heat Mass Transf 35:513–523

Han JC, Ou S, Park JS, Lei CK (1989) Augmented heat transfer in rectangular channels of narrow aspect ratios with rib turbulators. Int J Heat Mass Transf 32:1619–1630

Han JC, Zhang YM, Lee CP (1991) Augmented heat transfer in square channels with parallel, crossed, and V-shaped angled ribs. J Heat Transf 113:590–596

Han JC, Huang JJ, Lee CP (1993) Augmented heat transfer in square channels with wedge shaped and delta-shaped turbulence promoters. J Enhanc Heat Transf 1:37–52

Han JC, Huh M (2010) Recent studies in turbine blade internal cooling. Heat Transf Res 41:803–828

Han J, Glicksman L, Rohsenow W (1978) An investigation of heat transfer and friction for rib-roughened surfaces. Int J Heat Mass Transf 21(8):1143–1156

Han JC, Dutta S, Ekkad S (2000) Gas turbine heat transfer and cooling technology. Taylor & Francis, New York

Hans VS, Saini RP, Saini JS (2010) Heat transfer and friction factor correlations for a solar air heater duct roughened artificially with multiple V-ribs. Sol Energy 84:898–911

Herman C, Kang E (2001) Comparative evaluation of three heat transfer enhancement strategies in a grooved channel. Heat Mass Transf 37(6):563–575

Herman C, Kang E (2002) Heat transfer enhancement in a grooved channel with curved vanes. Int J Heat Mass Transf 45(18):3741–3757

Hijikata K, Ishiguro H, Mori Y (1987) Heat transfer augmentation in a pipe flow with smooth cascade turbulence promoters and its application to energy conversion. In: Yang WJ, Mori Y (eds) Heat transfer in high technology and power engineering. Hemisphere, New York, pp 368–397

Hishida M, Takase K (1987) Heat transfer coefficient of the ribbed surface. In: Proceedings of the third ASME/SME joint thermal engineering conference, vol 3, pp 103–110

Huang K, Wan J, Chen C, Mao D, Li Y (2013) Experiments investigation of the effects of surface roughness on laminar flow in macro tubes. Exp Thermal Fluid Sci 45:243–248

Hudina M (1979) Evaluation of heat transfer performances of rough surfaces from experimental investigation in annular channels. Int J Heat Mass Transf 22:1381–1392

Hwang J-J (1998) Heat transfer-friction characteristic comparison in rectangular ducts with slit and solid ribs mounted on one wall. J Heat Transf 120:709–716

Hwang J-J, Liou T-M (1994) Augmented heat transfer in a rectangular channel with permeable ribs mounted on the wall. J Heat Transf 116:912–920

Hwang J-J, Liou T-M (1995) Heat transfer and friction in a low-aspect-ratio rectangular channel with staggered perforated ribs on two opposite walls. J Heat Transf 117:843–850

Jaurker AR, Saini JS, Gandhi BK (2006) Heat transfer and friction characteristics of rectangular solar air heater duct using rib-grooved artificial roughness. Sol Energy 80:895–897

Jia R, Sunden B, Faghri M (2005) Computational analysis of heat transfer enhancement in square ducts with v-shaped ribs: turbine blade cooling. ASME J Heat Transf 127(4):425–433

Kamali R, Binesh AR (2008) The importance of rib shape effects on the local heat transfer and flow friction characteristics of square ducts with ribbed internal surfaces. Int Commun Heat Mass Transf 35(8):1032–1040

Kang M-G (2001) Diameter effects on nucleate pool boiling for a vertical tube. J Heat Transf 123:400–404

Kanoun M, Baccar M, Mseddi M (2011) Computational analysis of flow and heat transfer in passages with attached and detached rib arrays. J Enhanc Heat Transf 18(2):167–176

Karmare SV, Tikekar AN (2007) Heat transfer and friction factor correlation for artificially roughened duct with metal grit ribs. Int J Heat Mass Transf 50:4342–4351

Karwa R, Solanki SC, Saini JS (1999) Heat transfer coefficient and friction factor correlations for the transitional flow regime in rib-roughened rectangular ducts. Int J Heat Mass Transf 42:1597–1615

Karwa R (2003) Experimental studies of augmented heat transfer and friction in asymmetrically heated rectangular ducts with ribs on the heated wall in transverse inclined, V-continuous and V-discrete pattern. Int Commun Heat Mass Transf 30(2):241–250

Khalid A, Xie G, Sunden B (2016) Numerical simulations of flow structure and turbulent heat transfer in a square ribbed channel with varying rib pitch ratio. J Enhanc Heat Transf 23(2):155–174

Kim NH (2015) Single-phase pressure drop and heat transfer measurements of turbulent flow inside helically dimpled tubes. J Enhanc Heat Transf 22(4):345–363

Kong YQ, Yang LJ, Du XZ, Yang YP (2016) Air-side flow and heat transfer characteristics of flat and slotted finned tube bundles with various tube pitches. Int J Heat Mass Transf 99:357–371

Kukreja RT, Lau SC, McMillan RD (1993) Local heat/mass transfer distribution in a square channel with full and V-shaped ribs. Int J Heat Mass Transf 36:2013–2020

Kumar A, Bhagoria JL, Sarviya RM (2008) Heat transfer enhancement in channel of solar air collector by using discrete W-shaped artificial roughened absorber. In: Proc. 19th national and 8th ISHMT-ASME heat and mass transfer conference

Kumar A, Saini RP, Saini JS (2013) Development of correlations for Nusselt number and friction factor for solar air heater with roughened duct having multi V-shaped with gap rib as artificial roughness. Renew Energy 58:151–163

Kumar A, Saini RP, Saini JS (2014) A review of thermohydraulic performance of artificially roughened solar air heaters. Renew Sust Energ Rev 37:100–122

Kumar R, Judd RL (1970) Heat transfer with coiled wire turbulence promoters. Can J Chem Eng 48:378–383

Kumar S, Kothiyal AD, Bisht MS, Kumar A (2019) Effect of nanofluid flow and protrusion ribs on performance in square channels: an experimental investigation. J Enhanc Heat Transf 26 (1):75–100

Kumbhar DG, Sane NK (2015) Exploring heat transfer and friction factor performance of a dimpled tube equipped with regularly spaced twisted tape inserts. Procedia Eng 127:1142–1149

Kuwahara H, Takahashi K, Yanagida T, Nakayama W, Hzgimoto S, Oizumi K (1989) Method of producing a heat transfer tube for single-phase flow. U.S. patent 4,794,775, January 3

Lanjewar A, Bhagoria JL, Sarviya RM (2011) Experimental study of augmented heat transfer and friction in solar air heater with different orientations of W-rib roughness. Exp Thermal Fluid Sci 35:986–995

Layek A, Saini JS, Solanki SC (2006) Second law optimization of a solar air heater having chamfered rib-groove roughness on absorber plate. Renew Energy 32:1967–1980

Lee CK, Abdel-Moneim SA (2001) Computational analysis of heat transfer in turbulent flow past a horizontal surface with 2-D ribs. Int Commun Heat Mass Transf 28(2):161–170

Lee YO, Ahn J, Lee JS (2008) Effects of dimple depth and Reynolds number on the turbulent heat transfer in a dimpled channel. Prog Comput Fluid Dyn 8:432–438

Lee YO, Ahn J, Kim J, Lee JS (2012) Effect of dimple arrangements on the turbulent heat transfer in a dimpled channel. J Enhanc Heat Transf 19(4):359–367

Lewis MJ (1974) Roughness functions, the thermohydraulic performance of rough surfaces and the Hall transformation—an overview. Int J Heat Mass Transf 17:809–814

Li S, Xie G, Zhang W, Sunden B (2012) Numerical predictions of pressure drop and heat transfer in a blade internal cooling passage with continuous truncated ribs. Heat Transf Res 43:573–590

Liou TM, Hwang JJ (1993) Effect of ridge shapes on turbulent heat transfer and friction in a rectangular channel. Int J Heat Mass Transf 36:931–940

Liou T-M, Hwang J-J, Chen S-H (1993) Simulation and measurement of enhanced turbulent heat transfer in a channel with periodic ribs on one principal wall. Int J Heat Mass Transf 36:507–517

Ligrani PM, Oliveira MM, Blaskovich T (2003) Comparison of heat transfer augmentation techniques. AIAA J 41(3):337–362

Ligrani PM, Mahmood GI, Harrison JL, Clayton CM, Nelson DL (2001) Flow structure and local Nusselt number variations in a channel with dimples and protrusions on opposite walls. Int J Heat Mass Transf 44:4413–4425

Liu J, Song Y, Xie G, Sunden B (2015) Numerical modeling flow and heat transfer in dimpled cooling channels with secondary hemispherical protrusions. Energy 79:1–19

Luo J, Razinsky EH (2009) Analysis of turbulent flow in 180 deg turning ducts with and without guide vanes. ASME J Turbomach 131(2):021011

Mahmood GI, Ligrani PM (2002) Heat transfer in a dimpled channel: combined influences of aspect ratio, temperature ratio, Reynolds number, and flow structure. Int J Heat Mass Transf 45:2011–2020

Mahmood GI, Hill ML, Nelson DL, Ligrani PM, Moon HK, Glezer B (2001a) Local heat transfer and flow structure on and above a dimpled surface in a channel. J Turbomachinery 123:115–123

Mahmood GI, Sabbagh MZ, Ligrani PM (2001b) Heat transfer in a channel with dimples and protrusions on opposite walls. J Thennophys Heat Transf 15:275–283

Maubach K (1972) Rough annulus pressure drop—interpretation of experiments and recalculation for square ribs. Int J Heat Mass Transf 15:2489–2498

McLain CD (1975) Process for preparing heat exchanger tube. U.S. patent 1,906,605, issued to Olin Corp

Mehta MH, Raja Rao M (1979) Heat transfer and friction characteristics of spirally enhanced tubes for horizontal condensers. In: Chenoweth JM, et al (eds) Advances in enhanced heat transfer, ASME Symp. ASME, New York, pp 11–22

Mehta MH, Raja Rao M (1988) Analysis mid correlation of turbulent flow heat transfer and friction coefficients in spirally corrugated tubes for steam condenser application. In: Proceedings of the 1988 national heat transfer for conference, HTD-96, vol 3, pp 307–312

Meyer L (1980) Turbulent flow in a plane channel having one or two rough walls. Int J Heat Mass Transf 23:591–608

Mittal MK, Varun, Saini RP, Singal SK (2007) Effective efficiency of solar air heaters having different types of roughness elements on absorber plate. Energy 32:739–745

Momin AME, Saini JS, Solanki SC (2002) Heat transfer and friction in solar air heater duct with V-shaped rib roughness on absorber plate. Int J Heat Mass Transf 45:3383–3396

Muluwork KB (2000) Investigations on fluid flow and heat transfer in roughened absorber solar heaters. Ph.D. dissertation, Indian Institute of Technology Roorkee, Roorkee, Uttarakhand India

Murata A, Mochizuki S (2001) Comparison between laminar and turbulent heat transfer in a stationary square duct with transverse or angled rib turbulators. Int J Heat Mass Transf 44 (6):1127–1141

Nakayama W, Takahashi K, Daikoku T (1983) Spiral ribbing to enhance single-phase heat transfer inside tubes. In: Proceedings of the ASME-JSME thermal engineering joint conference, Honolulu, HI, vol 1. ASME, New York, pp 503–510

Naphon P (2007) Heat transfer characteristics and pressure drop in channel with v corrugated upper and lower plates. Energy Convers Manag 48(5):1516–1524

Naphon P, Nuchjapo M, Kurujareon J (2006) Tube side heat transfer coefficient and friction factor characteristics of horizontal tubes with helical rib. Energy Convers Manag 47:3031–3044

Newson IH, Hodgson TD (1973) The development of enhanced heat transfer condenser tubing. Desalination 14:291–323

Ooi A, Iaccarino G, Durbin PA, Behnia M (2002) Reynolds averaged simulation of flow and heat transfer in ribbed ducts. Int J Heat Fluid Flow 23:750–757

Ortiz L, Guerrero A, Arana C, Mendez R (2008) Heat transfer enhancement in a horizontal channel by the addition of curved deflectors. Int J Heat Mass Transf 51(15–16):3972–3984

Pal SK, Bhattacharyya S (2018) Enhanced heat transfer of cu-water nanofluid in a channel with wall mounted blunt ribs. J Enhanc Heat Transf 25(1):61–78

Pawar CB, Aharwal KR, Chaube A (2009) Heat transfer and fluid flow characteristics of rib-groove roughened solar air heater ducts. Indian J Sci Technol 2(11):50–54

Peng X, Peterson G (1996) Convective heat transfer and flow friction for water flow in microchannel structures. Int J Heat Mass Transf 39(12):2599–2608

Perng SW, Wu HW (2013) Heat transfer enhancement for turbulent mixed convection in reciprocating channels by various rib installations. J Enhanc Heat Transf 20(2):95–114

Prasad BN (2013) Thermal performance of artificially roughened solar air heaters. Sol Energy 91:59–67

Prasad K, Mulick SC (1983) Heat transfer characteristics of a solar air heater used for drying purposes. Appl Energy 13:83–93

Prasad BN, Saini JS (1988) Effect of artificial roughness on heat transfer and friction factor in a solar air heater. Sol Energy 41:555–560

Prasad BN, Behura AK, Prasad L (2014) Fluid flow and heat transfer analysis for heat transfer enhancement in three sided artificially roughened solar air heater. Sol Energy 105:27–35

Rabas TJ, Bergles AE, Moen DL (1988) Heat transfer and pressure drop correlations for spirally grooved (rope) tubes used in surface condensers and multistage flash evaporators. In: Augmentation of heat transfer in energy systems, ASME Symp. HTD, vol 52, pp 693–704

Rabas TJ, Thors P, Webb RL, Kim N-H (1993) Influence of roughness shape and spacing on the performance of three-dimensional helically dimpled tubes. J Enhanc Heat Transf 1:53–64

Raja Rao M (1988) Heat transfer and friction correlations for turbulent flow of water and viscous non Newtonian fluids in single-start spirally corrugated tubes. In: Proceedings of the 1988 national heat transfer conference HTD-96, vol 1, pp 677–683

Rau G, Cakan M, Moeller D, Arts T (1998) The effect of periodic ribs on the local aerodynamic and heat transfer performance of a straight cooling channel. ASME J Turbomach 120(2):368–375

Ravigururajan TS, Bergles AE (1985) General correlations for pressure drop and heat transfer for single-phase turbulent flow in internally ribbed tubes. In: Augmentation of heat transfer in energy systems, ASME Symp. HTD, vol 52, pp 9–20

Ravigururajan TS, Bergles AE (1996) Development and verification of general correlations for pressure drop and heat transfer in single-phase turbulent flow in enhanced tubes. Exp Therm Fluid Sci 13:55–70

Saha AK, Acharya S (2005) Flow and heat transfer in an internally ribbed duct with rotation: an assessment of large eddy simulations and unsteady Reynolds-averaged Navier–Stokes simulations. ASME J Turbomach 127(2):306–320

Sahu MM, Bhagoria JL (2005) Augmentation of heat transfer coefficient by using 90° broken transverse ribs on absorber plate of solar air heater. Renew Energy 30:2057–2063

Sahu MK, Prasad RK (2016) A review of the thermal and hydrodynamic performance of solar air heater with roughened absorber plates. J Enhanc Heat Transf 23(1):47–89

Saidi A, Sunden B (2000) Numerical simulation of turbulent convective heat transfer in square ribbed ducts. Numer Heat Transf 38:67–88

Saini RP, Saini JS (1997) Heat transfer and friction factor correlations for artificially roughened ducts with expanded metal mesh as roughened element. Int J Heat Mass Transf 40:973–986

Saini SK, Saini RP (2008) Development of correlations for Nusselt number and friction factor for solar air heater with roughened duct having arc-shaped wire as artificial roughness. Sol Energy 82:1118–1130

Saini RP, Verma J (2008) Heat transfer and friction factor correlations for a duct having dimple shaped artificial roughness for solar air heaters. Energy 33:1277–1287

Schüler M, Zehnder F, Weigand B, von Wolfersdorf J, Neumann SO (2011) The effect of turning vanes on pressure loss and heat transfer of a ribbed rectangular two-pass internal cooling channel. ASME J Turbomach 133(2):021017

Sethi M, Varun, Thakur NS (2012) Correlations for solar air heater duct with dimpled shape roughness elements on absorber plate. Sol Energy 86:2852–2861

Sethumadhavan R, Raja Rao M (1983) Turbulent flow heat transfer and fluid friction in helical wire coil inserted tubes. Int J Heat Mass Transf 26:1833–1845

Sethumadhavan R, Raja Rao M (1986) Turbulent flow friction and heat transfer characteristics of single- and multi-start spirally enhanced tubes. J Heat Transf 108:55–61

Sheriff N, Gumley P (1966) Heat transfer and friction properties of surfaces with discrete roughness. Int J Heat Mass Transf 9:1297–1320

Singh S, Chander S, Saini JS (2011) Heat transfer and friction factor correlations of solar air heater ducts artificially roughened with discrete V-down ribs. Energy 36:5053–5064

Slanciauskas A (2001) Two friendly rules for the turbulent heat transfer enhancement. Int J Heat Mass Transf 44:2155–2161

Sui Y, Teo C, Lee P, Chew Y, Shu C (2010) Fluid flow and heat transfer in wavy microchannels. Int J Heat Mass Transf 53(13):2760–2772

Suresh S, Chandrasekar M, Chandrasekar S (2001) Experimental studies on heat transfer and friction factor characteristics of CuO/water nanofluid under turbulent flow in a helically dimpled tube. Exp Thermal Fluid Sci 35:542–549

Takahashi K, Nakayama W, Kuwahara H (1988) Enhancement of forced convective heat transfer in tubes having three-dimensional spiral ribs. Heat Transf Jpn Res 17(4):12–28

Tanasawa I, Nishio S, Takano K, Tado M (1983) Enhancement of forced convection heat transfer in rectangular channel using turbulence promoters. In: Mori Y, Tanasawa I (eds) ASME-JSME thermal engineering joint conference, vol 1. ASME, New York, pp 395–402

Tanasawa I, Nishio S, Takano K, Miyazaki H (1985) Augmentation of forced convection heat transfer using novel rib-type turbulence promoters. Research on Effective use of Thermal Energy, The Ministry of Education

Tanda G (2004) Heat transfer in rectangular channels with transverse and V-shaped broken ribs. Int J Heat Mass Transf 47:229–243

Tanda G (2016) Performance of solar air heater ducts with different types of ribs on the absorber plate. Energy 36:6651–6660

Taslim ME, Spring SD (1988) An experimental investigation of heat transfer coefficients and friction factors in passages of different aspect ratios roughened with 45 degree turbulators. In: Proceedings of the 1988 national heat transfer conference, HTD-96, vol 1, pp 661–668

Taslim ME, Spring SD (1994) Effects of turbulator profile and spacing on heat transfer and friction in a channel. J Thermophys Heat Transf 8:555–562

Taslim ME, Setayeshgar L, Spring SD (2001) An experimental evaluation of advanced leading edge impingement cooling concepts. Int J Turbomach 123(1):147–153

Thianpong C, Eiamsa-ard P, Wongcharee K, Eiamsa-ard S (2009) Compound heat transfer enhancement of a dimpled tube with a twisted tape swirl generator. Int Commun Heat Mass Transf 36:698–704

Thors P, Clevinger NR, Campbell BJ, Tyler JT (1997) Heat transfer tubes and methods of fabrication thereof. U.S. patent 5,697,430, December 16

Thors P (2004) Personal communication

Varun, Saini RP, Singal SK (2007) A review on roughness geometry used in solar air heaters. Sol Energy 81:1340–1350

Varun, Saini RP, Singal SK (2008) Investigation of thermal performance of solar air heater having roughness elements as a combination of inclined and transverse ribs on absorber plate. Renew Energy 133:1398–1405

Verma SK, Prasad BN (2000) Investigation for the optimal thermohydraulic performance of artificially roughened solar air heaters. Renew Energy 20:19–36

Vicente PG, Garcia A, Viedma A (2002) Heat transfer and pressure drop for low Reynolds turbulent flow in helically dimpled tubes. Int J Heat Mass Transf 45:543–553

Viswanathan AK, Tafti DK (2006) A comparative study of DES and URANS for flow prediction in a two-pass internal cooling duct. ASME J Fluids Eng 128(6):1136–1345

Vyas S, Manglik RM, Milind AJ (2010) Visualization and characterization of a lateral swirl flow structure in sinusoidal corrugated-plate channels. J Flow Visual Image Process 17(4):281–296

Wang CC, Chen CK (2002) Forced convection in a wavy-wall channel. Int J Heat Mass Transf 45 (12):2587–2595

Wang L, Sunden B (2005) Experimental investigation of local heat transfer in square duct with continuous and truncated ribs. Exp Heat Transf 18:179–197

Webb RL, Kim NH (2005) Principles of enhanced heat transfer. Taylor & Francis, New York

Webb RL, Eckert ERG (1972) Application of rough surfaces to heat exchanger design. Int J Heat Mass Transf 15:1647–1658

Webb RL, Eckert ERG, Goldstein RJ (1971) Heat transfer and friction in tubes with repeated rib roughness. Int J Heat Mass Transf 14:601–617

Webb RL, Narayanamurthy R, Thors P (2000) Heat transfer and friction characteristics of internal helical-rib roughness. J Heat Transf 122:134–142

Wei X, Joshi Y, Ligrani P (2007) Numerical simulation of laminar flow and heat transfer inside a microchannel with one dimpled surface. ASME J Electron Packag 129(1):63–70

White L, Wilkie D (1970) The heat transfer and pressure loss characteristics of some multi-start ribbed surfaces. In: Augmentation of convective heat and mass transfer. ASME, New York, pp 55–62

Withada J, Boonloi A (2014) Effects of blockage ratio and pitch ratio on thermal performance in a square channel with 301 double V-affles. Case Stud Therm Eng 4:118–128

Withada J, Suwannapan S, Promvonge P (2011) Numerical study of laminar heat transfer in baffled square channel with various pitches. Energy Procedia 9:630–642

Wilkie D (1966) Forced convection heat transfer from surfaces roughened by transverse ribs. In: Third international heat transfer conference, vol 1, pp 1–19

Wilkie D, Cowan M, Burnett P, Burgoyne T (1967) Friction factor measurements in a rectangular channel with walls of identical and non-identical roughness. Int J Heat Mass Transf 10:611–621

Williams F, Pirie MAM, Warburton C (1970) Heat transfer from surfaces roughened by ribs. In: Augmentation of convective heat and mass transfer. ASME, New York, pp 55–62

Withers JG (1980a) Tube-side heat transfer and pressure drop for tubes having helical internal ridging with turbulent/transitional flow of single-phase fluid. Part 1: single-helix ridging. Heat Transf Eng 2(1):48–58

Withers JG (1980b) Tube-side heat transfer and pressure drop for tubes having helical internal ridging with turbulent/transitional flow of single-phase fluid. Part 2: multiple-helix ridging. Heat Transf Eng 2(2):43–50

Wongcharee K, Changcharoen W, Eiamsa-ard S (2011) Numerical investigation of flow friction and heat transfer in a channel with various shaped ribs mounted on two opposite ribbed walls. Int J Chem React Eng 9:26

Wright LM, Han JC (2014) Heat transfer enhancement for turbine blade internal cooling. J Enhanc Heat Transf 21(2–3):111–140

Xie GN, Zheng SF, Sunden B, Zhang WH (2013) A numerical investigation of flow structure and heat transfer enhancement in square ribbed channels with differently positioned deflectors. J Enhanc Heat Transf 20(3):195–212

Yadav S, Kaushal M, Varun (2013) Siddhartha Nusselt number and friction factor correlations for solar air heater duct having protrusions as roughness elements on absorber plate. Exp Thermal Fluid Sci 44:34–41

Zhang YF, Li FY, Liang ZM (1991) Heat transfer in spiral-coil-inserted tubes and its application. In: Ebadian MA, Pepper DW, Diller T (eds) Advances in heat transfer augmentation, ASME Symp. HTD, vol 169, pp 31–36

Zukauskas AA, Ulinskas RV (1983) Surface roughness as means of heat transfer augmentation for banks of tubes in crossflow. In: Taborek J, Hewitt GP, Afgan N (eds) Heat exchangers: theory and practice. Hemisphere, Washington, DC, pp 311–321

Zukauskas AA, Ulinskas RV (1988) Heat transfer in tube banks in crossflow. Hemisphere, New York, pp 94–118

Chapter 7
Compound Techniques

Saha and co-authors have presented intensive data on the performance of combination of passive inserts for heat transfer enhancement in flow through circular, square and rectangular channels. Rout and Saha (2013) studied the heat transfer and pressure drop characteristics in flow through pipe using a compound technique which consists of wire coil and helical screw tape inserts. The correlations for Nusselt number and friction factor have also been presented using log regression method. In the development of correlations for Nusselt number and friction factor, the effect of different parameters has been considered: geometrical parameters of the wire coil and helical screw tape inserts, thermal entrance length (Graetz number), buoyancy forces (Rayleigh number) at low *Re* and swirl and gravity forces at high *Re*.

The wire coil has been used to mix the gross flow while the helical screw causes the flow to spiral along the tube length, thus increasing the effective length of flow. They have observed that the use of wire coil results in the disturbance of hydrodynamic boundary layer than the thermal boundary layer. Also, a higher momentum loss due to mixing has been observed. However, the frequency of separation and reattachment of the thermal boundary layer was increased resulting in Nusselt number. This results in greater increase in heat transfer over the increase in pressure drop, making the wire coil and helical screw tape insert an effective technique for heat transfer enhancement in laminar flow through pipe.

Saha and Dutta (2001) studied the performance of regularly spaced twisted tape elements and twisted tapes with gradually decreasing pitch. They observed that the performance of regularly spaced twisted tape with multiple twists was quite similar to that with a single twist, in the given range of Reynolds number. Also, they concluded that uniform-pitch twisted tapes performed better than those having gradually decreasing pitch. Improved heat transfer performance, using regularly spaced twisted tape over that with full-length twisted tape, has been observed by Saha et al. (1989) and Date and Saha (1990).

© The Author(s), under exclusive license to Springer Nature Switzerland AG 2020 159
S. K. Saha et al., *Insert Devices and Integral Roughness in Heat Transfer Enhancement*, SpringerBriefs in Applied Sciences and Technology,
https://doi.org/10.1007/978-3-030-20776-2_7

Saha (2013) presented the performance of compound technique for laminar flow through a pipe with integral helical corrugations fitted with helical screw tape insert. The application of this compound technique in the designing of tubes of parabolic trough solar collectors has been proposed. The integral helical corrugation channels are sinusoidal channels. The increase in Nusselt number and friction factor with increase in corrugation angle has been reported.

The performance of integral transverse ribs along with centre-cleared twisted tape for heat transfer enhancement in a circular channel in laminar flow regime has been studied by Bhattacharyya et al. (2013). They observed an increase in Nu and f for increase in central clearance (c) till $c = 0.4$ mm, after which the change in Nu and f with central clearance becomes negligible. About 75–135% increase in Nusselt number has been noted while the increase in pressure drop was around 20–35%.

Saha (2010a) studied the combined effect of transverse ribs and wire coil inserts in rectangular and square ducts. The viscous oil ($195 < Pr < 525$) had been used as the working fluid. The transverse ribs were provided on the opposite sides of the duct. They reported almost 50% increase in heat duty for the compound technique over that for individual inserts based on constant pumping power performance evaluation criterion. Also, the reduction in pressure dropped up to 40% using combined insert geometry based on constant heat duty performance criteria.

Saha (2010b) investigated the thermohydraulic performance of axial corrugation roughness and twisted tapes for turbulent flow in ducts. The twisted tapes with and without oblique teeth cut on them had been considered. The turbulent flow of air with $10,000 < Re < 100,000$ had been considered. The heat transfer investigations were conducted in stainless steel ducts with electrical heating under constant wall heat flux boundary condition. The pressure drop investigations, on the other hand, had been carried out in acrylic channels. The effect of duct aspect ratio, corrugation angle, corrugation pitch, twist ratio, space ratio, length of the twisted tape, oblique tooth angle and horizontal tooth length on heat transfer and pressure drop characteristics was presented. They observed that the combination of axial corrugation roughness and twisted tape with oblique teeth resulted in better heat transfer performance as compared to that of the combination of corrugation roughness and twisted tape without oblique teeth.

Bhattacharyya and Saha (2012) presented the performance of integral helical rib roughness fitted with centre-clearance twisted tape along with correlations. The flow through circular channel in laminar flow regime had been considered. Saha and Saha (2013b) worked with the integral helical rib roughness along with helical screw tapes. The combination of integral helical rib roughness and wavy strip inserts was studied by Saha and Saha (2013a) for laminar flow through circular tube. Pal and Saha (2014) studied the tube having spiral corrugation roughness fitted with twisted tape having oblique teeth.

Pal and Saha (2015) worked with tube having spiral ribs and twisted tapes. Combination of helical screw tape with oblique teeth and wire coil was considered by Roy and Saha (2015), for heat transfer and pressure drop investigations of laminar flow through a circular channel. Saha et al. (2012b) considered circular tube with axial corrugations fitted with helical screw tape inserts for heat transfer augmentation in laminar flow. Saha et al. (2012b) studied axial rib roughness with helical screw tape inserts. Pal and Saha (2010) dealt with transverse ribs and twisted tapes with and without oblique teeth in square and rectangular ducts.

Saha and Langille (2002) studied the thermohydraulic performance of laminar flow through a circular tube using longitudinal strip inserts for heat transfer enhancement under constant wall heat flux boundary condition. Saha and Mallick (2005) studied the heat transfer performance of laminar flow in rectangular and square ducts with twisted tape inserts. Saha et al. (2012a) used wire coil inserts in combination with centre-cleared twisted tape. It has been concluded from all the above investigations that the performance of combination of various inserts is found to be better than that of individual inserts used for heat transfer augmentation.

Zimparov et al. (2006) studied the effect of compound technique on the performance of heat transfer characteristics of rough surfaces with tube inserts. They compared the experimental works which were done until now by the passive combined technique in a single-phase flow like twisted tape inserts combined with rough tube (sand grain roughness), internally grooved rough tube, three-dimensional extended internal surfaces and corrugated tube. An internally grooved tube with wire coil was also considered. They evaluated the performance characteristics of rough surfaces with insert by using entropy generation method. The influence of geometric parameters such as roughness height-to-diameter ratio (e/D_i), relative pitch (p/e), relative helix angle β_* and others was also studied. Figure 7.1 shows different types of rough surfaces (Fig. 7.1a–f) and a twisted tape (Fig. 7.1g).

Bergles et al. (1969), Usui et al. (1986), Zimparov (2001, 2002), Hasim et al. (2003a, b) and Liao and Xin (2000) investigated heat transfer compound techniques such as (a) combination of rough tube with a twisted tape, (b) internally, (c) grooved tube with a twisted tape insert, (d) corrugated tubes with a twisted tape, (e) helically ribbed tube with a twisted tape insert, (f) extended surfaces and twisted tape inserts shown in Fig. 7.1. Table 7.1 shows the geometrical parameters of the tested tubes. Table 7.2 shows the geometrical parameters of the tube inserts (Hasim et al. 2003a, b). Table 7.3 shows the geometrical parameters of the 3-DIEST tubes (Liao and Xin 2000). Figure 7.2 shows the variation of friction versus Reynolds number. Figure 7.3 presents the results from the heat transfer studies carried out to obtain values for water-side heat transfer coefficients.

Zimparov et al. (2012) presented the effectiveness of using compound heat transfer technique consisting of corrugated tubes with twisted tapes for heat transfer augmentation in single-phase flow. They considered eight tubes with deep single-start spiral corrugations and twisted tape inserts. They observed that the effect of twisted tape pitch on friction factor became negligible when used in the corrugated tube. For deep corrugations, the friction factor was found to increase with the increase in the ratio of rib height and tube diameter (e/D). However, the friction factor decreased with increasing Reynolds number. The increase in heat transfer

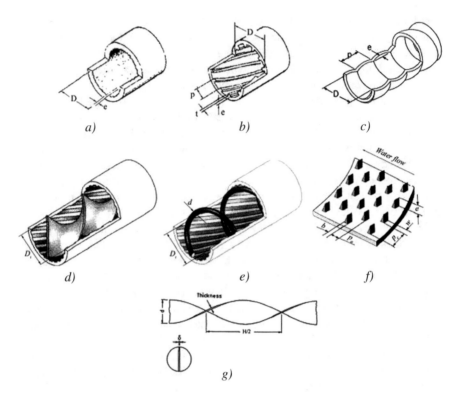

Fig. 7.1 Overall view of the compound techniques studied (Zimparov et al. 2006)

Table 7.1 Geometrical parameters of the tested tubes (Zimparov et al. 2006)

No.	D_i (mm)	e (mm)	p (mm)	β (deg)	e/D_i	p/e	β_*	H/D_i	NDC
1	6.35	0.076	–	–	0.012	–	–	12.32	–
2	6.35	0.076	–	–	0.012	–	–	8.34	–
3	6.35	0.076	–	–	0.012	–	–	5.10	–
4	14.17	0.309	1.04	30.0	0.022	3.37	0.333	8.40	–
5	14.17	0.309	1.04	30.0	0.022	3.37	0.333	8.40	–
6	13.90	0.312	5.76	82.4	0.022	18.5	0.916	4.68	–
7	13.15	0.593	5.06	83.0	0.045	8.50	0.922	4.94	–
8	13.51	0.767	5.19	83.0	0.057	6.67	0.922	5.83	–
9	19.20	0.350	3.02	11.0	0.018	8.62	0.122	4.16	0.13
10	19.20	0.350	3.02	11.0	0.018	8.62	0.122	5.21	0.02
11	19.50	0.850	3.06	–	0.044	3.60	–	1.03	–
12	19.20	0.380	3.02	11.0	0.020	7.95	0.122	1.82	0.04
13	19.50	0.850	3.06	–	0.044	3.60	–	1.03	0.18
14	19.20	0.380	3.02	11.0	0.020	7.95	0.122	1.82	0.21

Table 7.2 Geometrical parameters of the tube inserts (Hasim et al. 2003a, b)

	H (mm)	D_d (mm)	d (mm)	δ (mm)	Helical direction
Coil A	20.0	17.8	2.0	–	Clockwise
Coil B	35.0	17.8	2.0	–	Clockwise
Twisted tape A	80.0	16.2	–	1.0	Counterclockwise
Twisted tape B	100.0	18.2	–	1.0	Counterclockwise

Table 7.3 Geometrical parameters of the 3-DIEST tubes (Liao and Xin 2000)

No.	e/D_i	p_a/e	w/p_a	p_c/w	H/D_i
15	0.077	4.080	0.118	5.421	30.0
16	0.077	4.080	0.118	5.421	20.0
17	0.077	4.080	0.118	5.421	10.0

Fig. 7.2 Variation of friction factor f with Re (Zimparov et al. 2006)

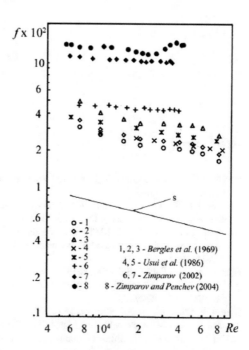

coefficient was observed with decrease in pitch of the twisted tape. They reported that the minimum heat transfer enhancement ratio was observed for tubes with $e/D = 0.057$ and $P/e = 9.5$. Promvonge (2008), Bharadwaj et al. (2009) and Zimparov et al. (2006) also worked with various compound techniques.

Fig. 7.3 Variation of
$NuPr^{-0.4}$ with Re
(Zimparov et al. 2006)

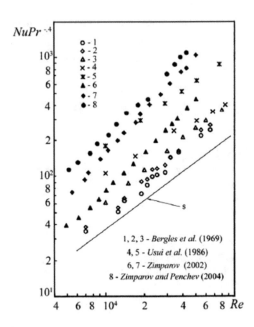

References

Bergles AE, Lee RA, Mikic BB (1969) Heat transfer in rough tubes with tape-generated swirl flow. J Heat Transf 91:443–445

Bharadwaj P, Khondge AD, Date AW (2009) Heat transfer and pressure drop in a spirally grooved tube with twisted tape insert. Int J Heat Mass Transf 52:1938–1944

Bhattacharyya S, Saha SK (2012) Thermohydraulics of laminar flow through a circular tube having integral helical rib roughness and fitted with centre-cleared twisted-tape. Exp Thermal Fluid Sci 42:154–162

Bhattacharyya S, Saha S, Saha SK (2013) Laminar flow heat transfer enhancement in a circular tube having integral transverse rib roughness and fitted with centre-cleared twisted-tape. Exp Thermal Fluid Sci 44:727–735

Date AW, Saha SK (1990) Numerical prediction of laminar flow in a tube fitted with regularly spaced twisted-tape elements. Int J Heat Fluid Flow 11(4):346–354

Hasim F, Yoshida M, Miyashita H (2003a) Compound heat transfer enhancement by a combination of ribbed tubes with wire coil inserts. J Chem Eng Japan 36(6):647–654

Hasim F, Yoshida M, Miyashita H (2003b) Compound heat transfer enhancement by a combination of helically ribbed tubes with twisted tape inserts. J Chem Eng Japan 36(9):1116–1122

Liao Q, Xin MD (2000) Augmentation of convective heat transfer inside tubes with three-dimensional internal extended surface and twisted tape inserts. Chem Eng J 78(2–3):95–105

Pal PK, Saha SK (2010) Thermal and friction characteristics of laminar flow through square and rectangular ducts with transverse ribs and twisted tapes with and without oblique teeth. J Enhanc Heat Transf 17(1):1–21

Pal PK, Saha SK (2014) Experimental investigation of laminar flow of viscous oil through a circular tube having integral spiral corrugation roughness and fitted with twisted tapes with oblique teeth. Exp Thermal Fluid Sci 57:301–309

Pal S, Saha SK (2015) Laminar fluid flow and heat transfer through a circular tube having spiral ribs and twisted tapes. Exp Thermal Fluid Sci 60:173–181

Promvonge P (2008) Thermal augmentation in circular tube with twisted tape and wire coil turbulators. Energy Convers Manag 49:2949–2955

Rout PK, Saha SK (2013) Laminar flow heat transfer and pressure drop in a circular tube having wire-coil and helical screw-tape inserts. J Heat Transf 135(2):021901

Roy S, Saha SK (2015) Thermal and friction characteristics of laminar flow through a circular duct having helical screw-tape with oblique teeth inserts and wire coil inserts. Exp Thermal Fluid Sci 68:733–743

Saha SK (2010a) Thermal and friction characteristics of laminar flow through rectangular and square ducts with transverse ribs and wire coil inserts. Exp Thermal Fluid Sci 34(1):63–72

Saha SK (2010b) Thermohydraulics of turbulent flow through rectangular and square ducts with axial corrugation roughness and twisted-tapes with and without oblique teeth. Exp Thermal Fluid Sci 34(6):744–752

Saha SK (2013) Thermohydraulics of laminar flow through a circular tube having integral helical corrugations and fitted with helical screw-tape insert. Chem Eng Commun 200(3):418–436

Saha SK, Dutta A (2001) Thermohydraulic study of laminar swirl flow through a circular tube fitted with twisted tapes. ASME J Heat Transf 123:417–427

Saha SK, Langille P (2002) Heat transfer and pressure drop characteristics of laminar flow through a circular tube with longitudinal strip inserts under uniform wall heat flux. J Heat Transf 124 (3):421–432

Saha SK, Mallick DN (2005) Heat transfer and pressure drop characteristics of laminar flow in rectangular and square plain ducts and ducts with twisted-tape inserts. J Heat Transf 127 (9):966–977

Saha S, Saha SK (2013a) Enhancement of heat transfer of laminar flow through a circular tube having integral helical rib roughness and fitted with wavy strip inserts. Exp Thermal Fluid Sci 50:107–113

Saha S, Saha SK (2013b) Enhancement of heat transfer of laminar flow of viscous oil through a circular tube having integral helical rib roughness and fitted with helical screw-tapes. Exp Thermal Fluid Sci 47:81–89

Saha SK, Gaitonde UN, Date AW (1989) Heat transfer and pressure drop characteristics of laminar flow in a circular tube fitted with regularly spaced twisted-tape elements. Exp Thermal Fluid Sci 2:310–322

Saha SK, Barman BK, Banerjee S (2012a) Heat transfer enhancement of laminar flow through a circular tube having wire coil inserts and fitted with center-cleared twisted tape. J Therm Sci Eng Appl 4(3):031003

Saha S, Bhattacharyya S, Dayanidhi GL (2012b) Enhancement of heat transfer of laminar flow of viscous oil through a circular tube having integral axial rib roughness and fitted with helical screw-tape inserts. Heat Transf Res 43(3):207–227

Usui N, Sano Y, Iwashita K, Isozaki A (1986) Enhancement of heat transfer by a combination of an internally grooved rough tube and a twisted tape. Int Chem Eng 26(1):97–104

Zimparov VD (2001) Enhancement of heat transfer by a combination of three-start spirally corrugated tubes with a twisted tape. Int J Heat Mass Transf 44(3):551–574

Zimparov VD (2002) Enhancement of heat transfer by a combination of a single-start spirally corrugated tubes with a twisted tape. Exp Thermal Fluid Sci 25:535–546

Zimparov V, Penchev PJ, Bergles AE (2006) Performance characteristics of some "rough surfaces" with tube inserts for single-phase flow. J Enhanc Heat Transf 13(2):117–137

Zimparov V, Petkov VM, Bergles AE (2012) Performance characteristics of deep corrugated tubes with twisted-tape inserts. J Enhanc Heat Transf 19(1):1–11

Chapter 8
Conclusions

The following conclusions may be drawn from the discussion made in this research monograph:

- Insert devices are not good competitor with internally finned tubes or roughness for turbulent flow.
- However, insert devices are good for laminar flow of viscous liquids.
- Integral roughness is not good for laminar flow since the roughness is too small to enhance a laminar flow.
- Insert devices are used to upgrade the performance of an existing heat exchanger having a plain inner tube surface
- Major insert devices are twisted tapes, wire coil inserts, extended surface inserts, mesh or brush inserts and displaced inserts
- Many design correlations are available for inserts, but this is not enough.
- More research on tangential swirl injection is needed.
- More information on performance characteristics with various insert devices is needed.
- Research on fouling with insert devices is necessary.
- Significant progress has been made in the design and manufacture of roughened surfaces. However, more advanced research is necessary.
- A properly designed rough surface must have good efficiency index.
- More and more advanced manufacturing techniques are likely to give better rough surfaces for enhancement.
- Geometrical similarity makes the design of rough surfaces more convenient.
- Although predictive methods have already made significant inroads, better predictive models are required for rough surfaces.
- Heat transfer surface fouling is a matter of concern for rough surfaces.

© The Author(s), under exclusive license to Springer Nature Switzerland AG 2020
S. K. Saha et al., *Insert Devices and Integral Roughness in Heat Transfer Enhancement*, SpringerBriefs in Applied Sciences and Technology,
https://doi.org/10.1007/978-3-030-20776-2_8

Additional References

Alamgholilou A, Esmaeilzadeh E (2012) Experimental investigation on hydrodynamics and heat transfer of fluid flow into channel for cooling of rectangular ribs by passive and EHD active enhancement methods. Exp Thermal Fluid Sci 38:61–73

Bell KJ, Mueller AC (eds) (1984) Engineering Data Book II. Wolverine Tube Corp., Decatur, AL

Bhatia RS, Webb RL (2001) Numerical study of turbulent flow and heat transfer in microfin tubes. Part I, model validation. J Enhanc Heat Transf 8:291–304

Chandra PR, Niland ME, Han JC (1997) Effect of rib profiles on turbulent channel flow heat transfer. J Turbomach 119:373–380

Chang SW, Yu KW, Lu MH (2005) Heat transfers in tubes fitted with single, twin and triple twisted tapes. Exp Heat Transf 18:279–294

Chiu Y, Jang J (2009) 3D numerical and experimental analysis for thermal–hydraulic characteristics of air flow inside a circular tube with different tube inserts. Appl Therm Eng 29:250–258

Cuangya L, Chuanyun G, Chaosu W, Jinshu H, Cun J (1991) Experimental investigation of transitional flow heat transfer of three-dimensional internally finned tubes. In: Ebadian MA, Pepper DW, Diller T (eds) Advances in heat in transfer, augmentation and mixed convection, ASME Symp., HTD, vol 169. ASME, New York, pp 45–48

Deakin AW, Honda H, Rudy TM (eds) (2001) Begell House, New York, pp 187–198

Eiamsa-ard S, Wongcharee K, Sripattanapipat S (2009) 3-D Numerical simulation of swirling flow and convective heat transfer in a circular tube induced by means of loose-fit twisted tapes. Int Commun Heat Mass Transf 36:947–955

Elshafei EAM, Safwat Mohamed M, Mansour H, Sakr M (2008) Experimental study of heat transfer in pulsating turbulent flow in a pipe. Int J Heat Fluid Flow 29:1029–1038

Good MC, Joubert PN (1968) The form drag of two-dimensional bluff plates immersed in turbulent boundary layers. J Fluid Mech 31:547–582

Hall WB (1962) Heat transfer in channels having rough and smooth surfaces. J Mech Eng Sci 4:287–291

Han JC, Glicksman LR, Rohsenow WM (1979) An investigation of heat transfer and friction for rib-roughened surfaces. Int J Heat Mass Transf 22:1587

Hay N, West PD (1975) Heat transfer in free swirl in flow in a pipe. J Heat Transf 97:411–416

Hitachi (1984) Hitachi high-performance heat-transfer tubes, Catalog EA-500. Hitachi Cable, Ltd, Tokyo, Japan

Hosni MH, Coleman HW, Taylor RP (1989) Measurement and calculation of surface roughness effects on turbulent flow and heat transfer, Report TFD-89-1, Mechanical and Nuclear Engineering Department, Mississippi State University

© The Author(s), under exclusive license to Springer Nature Switzerland AG 2020
S. K. Saha et al., *Insert Devices and Integral Roughness in Heat Transfer Enhancement*, SpringerBriefs in Applied Sciences and Technology,
https://doi.org/10.1007/978-3-030-20776-2

Hosni MH, Coleman HW, Taylor RP (1991) Measurements and calculations of rough-wall heat transfer in the turbulent boundary layer. Int J Heat Mass Transf 34:1067–1081

Ibrahim EZ (2011) Augmentation of laminar flow and heat transfer in flat tubes by means of helical screw-tape inserts. Energy Convers Manag 52:250–257

Inaba H, Ozaki K (1997) Heat transfer enhancement and flow-drag reduction of forced convection in circular tubes by means of wire coil insert. In: Shah RK, Bell KJ, Mochizuki S, Wadekar VW (eds) Proceedings of the international conference on compact heat exchangers for the process industries. Begell House Inc., New York, pp 445–452

Jaisankar S, Radhakrishnan TK, Sheeba KN (2008) Experimental studies on heat transfer and friction factor characteristics of forced circulation solar water heater system fitted with left–right twisted tapes. Int Energy J 9:1–5

Jaisankar S, Radhakrishnan TK, Sheeba KN (2009) Experimental studies on heat transfer and friction factor characteristics of thermosyphon solar water heater system fitted with spacer at the trailing edge of twisted tapes. Appl Therm Eng 29:1224–1231

Jaisankar S, Radhakrishnan TK, Sheeba KN (2011) Experimental studies on heat transfer and thermal performance characteristics of thermosyphon solar water heating system with helical and left–right twisted tapes. Energy Convers Manag 52:2048–2055

Karwa R, Solanki SC, Saini JS (2001) Thermo-hydraulic performance of solar air-heaters having integral chamfered rib roughness on absorber plates. Energy 26:161–176

Krishna SR, Pathipaka G, Sivashanmugam P (2008) Studies on heat transfer augmentation in a circular tube fitted with straight half twist left–right inserts in laminar flow. J Environ Res Dev 3:437–441

Krishna SR, Pathipaka G, Sivashanmugam P (2009) Heat transfer and pressure drop studies in a circular tube fitted with straight full twist. Exp Thermal Fluid Sci 33:431–438

Lewis MJ (1975) An elementary analysis predicting the momentum and heat transfer characteristics of a hydraulically rough surface. J Heat Transf 97:249–254

Li HM, Ye KS, Tan YK, Den SJ (1982) Investigation of tube-side flow visualization, friction factors and heat transfer characteristics of helical-ridging tubes. In: Proceedings 7th international heat transfer conference, vol 3. Hemisphere, Washington, DC, pp 75–80

Liou T-M, Chen S-H, Shih K-C (2002) Numerical simulation of turbulent flow field and heat transfer in a two-dimensional channel with periodic slit ribs. Int J Heat Mass Transf 45:4493–4505

Liu X, Jensen MK (2001) Geometry effects on turbulent flow and heat transfer in internally finned tubes. J Heat Transf 123:1035–1044

Mahmood GI (2001) Heat transfer and flow structure from dimples in an internal cooling passage. Ph.D. thesis, University of Utah, Salt Lake City

Moon H-K, O'Conell T, Glezer B (1999) Channel height effect on heat transfer and friction in a dimpled passage, ASME Paper 99-GT-163. ASME, New York

Muluwork KB, Solanki SC, Saini JS (2000) Study of heat transfer and friction in solar air heaters roughened with staggered discrete ribs. In: Proc. 4th ISHMT-ASME heat and mass transfer conf., Pune, India, pp 391–398

Murugesan P, Mayilsamy K, Suresh S, Srinivasan P (2009) Heat transfer and pressure drop characteristics of turbulent flow in a tube fitted with trapezoidal-cut twisted tape insert. Int J Acad Res 1:123–127

Murugesan P, Mayilsamy K, Suresh S (2010) Turbulent heat transfer and pressure drop in tube fitted with square-cut twisted tape. Fluid Flow Transp Phenom Chin J Chem Eng 18:609–617

Nakamura H, Tanaka M (1973) Cross-rifled vapor generating tube. U.S. patent 3,734,140

Obot NT, Das L, Rabas TJ (2001) Smooth- and enhanced-tube heat transfer and pressure drop. Part I. Effect of Prandtl number with air, water and glycol/water mixtures. In: Shah RK, Deakin AW, Honda H, Rudy TM (eds) Proceedings of the third international conference on compact heat exchangers and enhancement technology for the process industries. Begell House, New York, pp 259–264

Prasad RC, Brown MJ (1988) Effectiveness of wire-coil inserts in augmentation of convective heat transfer. In: Shah RK (ed) Experimental heat transfer; fluid mechanics and thermodynamics

Promvonge P (2008a) Thermal enhancement in a round tube with snail entry and coiled-wire inserts. Int Commun Heat Mass Transf 35:623–629

Promvonge P (2008b) Thermal performance in circular tube fitted with coiled square wires. Energy Convers Manag 49:980–987

Rahimia M, Shabanian SR, Alsairafi AA (2009) Experimental and CFD studies on heat transfer and friction factor characteristics of a tube equipped with modified twisted tape inserts. Chem Eng Process 48:762–770

Shabanian SR, Rahimi M, Shahhosseini M, Alsairafi AA (2011) CFD and experimental studies on heat transfer enhancement in an air cooler equipped with different tube inserts. Int Commun Heat Mass Transf 38:383–390

Shah RK, Bhatti MS (1987) Laminar convective heat transfer in ducts. In: Kakac S, Shah RK, Aung W (eds) Handbook of single phase heat transfer. Wiley, New York, pp 3–20

Shivkumar C, Raja Rao M (1988) Studies on compound augmentation of laminar flow heat transfer to generalized power law fluids in spirally corrugated tubes by means of twisted tape inserts. Proc Natl Heat Transf Conf ASME-HTD-ASME 96(1):685–691

Shome B, Jensen MK (1996) Experimental investigation of laminar flow and heat transfer in internally finned tubes. J Enhanc Heat Transf 4:53–70

Sivashanmugam P, Nagarajan PK (2007) Studies on heat transfer and friction factor characteristics of laminar flow through a circular tube fitted with right and left helical screw-tape inserts. Exp Thermal Fluid Sci 32:192–197

Sumitomo (1983) Technical data of Tred-fin. Sumitomo Light Metal Industries, Aichi, Japan

Thomas DG (1967) Enhancement of forced convection mass transfer coefficient using detached turbulence promoters. Ind Eng Chem Process Design Dev 6:385–390

Tiwari M, Saha SK (2015) Laminar flow through a circular tube having transverse ribs and twisted tapes. J Therm Sci Eng Appl 7(4):041009

Tu W, Xu Z, Zhu Y, Wang Y, Tang Y (2017) Boundary layer redevelops with mesh cylinder inserts for heat transfer enhancement. Int J Heat Mass Transf 109:147–156

Uttawar SB, Raja Rao M (1985) Augmentation of laminar flow beat transfer in tubes by means of wire coil inserts. J Heat Transf 105:930–935

Webb RL (1971) A critical evaluation of analytical and Reynolds analogy equations for turbulent heat and mass transfer in smooth tubes. Wiirme Stoffubertrag 4:197–204

Webb RL (1982) Performance cost effectiveness and water-side fouling considerations of enhanced tube heat exchangers for boiling service with tube-side water flow. Heat Transf Eng 3(3):84–98

Webb RL, Eckert ERG, Goldstein RJ (1972) Generalized heat transfer and friction correlations for tubes with repeated-rib roughness. Int J Heat Mass Transf 15:180–184

Wolfstein M (1988) The velocity and temperature distribution of one dimensional flow with turbulence augmentation and pressure gradient. Int J Heat Mass Transf 12:301–318

Wongcharee K, Eiamsa-ard S (2011) Heat transfer enhancement by twisted tapes with alternate-axes and triangular, rectangular and trapezoidal wings. Chem Eng Process Process Intensif 50:211–219

Yakut K, Sahin B (2004) The effects of vortex characteristics on performance of coiled wire turbulators used for heat transfer augmentation. Appl Therm Eng 24:2427–2438

Yorkshire (1982) Heat exchanger tubes: design data for horizontal rope tubes in steam condensers, Technical Memorandum 3. Yorkshire Imperial Metals, Ltd., Leeds, UK

Index

Printed in the United States
By Bookmasters